I0656059

Light Years from Tranquility

Other books in the Light Years series:

Light Years from Paradise: Einstein's Double-Take

Light Years from Tranquility

FRANK LEWANDOWSKI

RESOURCE *Publications* · Eugene, Oregon

LIGHT YEARS FROM TRANQUILITY

Copyright © 2011 Frank Lewandowski. All rights reserved. Except for brief quotations in critical publications or reviews, no part of this book may be reproduced in any manner without prior written permission from the publisher. Write: Permissions, Wipf and Stock Publishers, 199 W. 8th Ave., Suite 3, Eugene, OR 97401.

Resource Publications
An Imprint of Wipf and Stock Publishers
199 W. 8th Ave., Suite 3
Eugene, OR 97401

www.wipfandstock.com

ISBN 13: 978-1-61097-266-6

Manufactured in the U.S.A.

Dedicated to Almighty God who makes the impossible, possible.

Acknowledgments

Thanks to the numerous relatives, church brethren, friends, and coworkers who have shown me encouragement and support regarding my science fiction writing. I especially appreciate my wife Sandy, our daughter Holly and son Michael, my sister Mary Anne Stabile, Mike Read, Ben Martinez, Larisa Clopton, Jean Grunheid and Sarah Crespo. Thanks to Laniece Miller for sending me a star map so I could locate *Light Years From Tranquility* within interstellar geography.

It can sometimes seem like a long journey from story idea to published novel so the help of so many means a lot.

—Frank Lewandowski
December, 2010

Hostile Fleet from Space

FIVE TEMPORAL UNITS UNTIL the planet dies.

The vast space fleet advanced toward its destination, a small orange dot that was gradually growing larger as seen through the ship's ports. The target was the small settlement on the world's surface, the pitiful village containing sentient beings. Other than their own species, the aliens who inhabited the ships hated sentient beings. The fleet's objective was the same as always: to preserve and enhance its race's position as the dominant beings in that part of the galaxy. The aliens had previously sent scouting missions that had flown by the planet and studied the puny colony. It was of fairly recent origin and populated by a species that was new to this sector, creatures with soft, fleshy shells, body hair and only four limbs. Who were they to encroach on this part of space? And especially to plant their seed on a world that looked so much like the Home Planet. This desecration had to be punished. It was time of make an example of these interlopers before their kind spread to yet other star systems.

The invaders received first one then a second plea from the local inhabitants. The colony's residents were begging for mercy. The commanders ignored their cries. The plan was still on. It gave the predators twisted pleasure to get a hint of the anguish they were causing.

Two temporal units until the planet dies.

The armada, ten of thousands strong passed by the planet's moon. So massive was the swarm that its shadow covered much of the tiny orb.

The ships barely slowed long enough to destroy the tiny blip of a micro-colony on the lunar surface.

One temporal unit.

The ships entered into orbit around its target, encircling the ill-fated world like the rings that surrounded the Home Orb. The swarm moved in closer, tightening its death grip.

Time up. Planet dies.

A number of the ships set their sites on the colony. They opened fire, bathing the offenders in a blazing holocaust. Within moments, it was all over. A grey, smoking crater was all that remained where thousands had once lived. Their mission completed, the fleet headed back into deep space, its leaders confident they had vanquished all other intelligent life in this star system.

Or so they thought . . .

2

Mysterious Invaders

SEVERAL HOURS EARLIER, EDD Brawnley had sat in the observatory at the lunar base that served as the local sub-colony. A two-foot thick layer of ice composed the base's outermost walls, a precaution that shielded the residents from solar flares and cosmic rays. The main habitat was on the planet the moon circled. Edd was tall with broad shoulders, a thin waist and a blond crew cut. His job was to interpret feedback from the billions of nano satellites that formed an invisible halo around the moon, serving as the colony's eyes and ears in space. Today had been a slow work shift. The system that sorted through the immense amounts of data from the bots had apparently found little that might be of interest to humans. The astronomer gazed at the monitor that displayed a real-time view of the grey, pock marked lunar landscape, the blackness of space and the countless bright stars that studded the sky. The planet hung over the horizon like an orange marble. The sun was a distant ball of yellow-white fire. The human had gotten tired of gazing at the scenery and had been spending some time playing simple on line games. Abruptly, a flashing red screen materialized two feet from his face. "Finally, something to do," he thought.

He raised his eyebrows. A number of the bots had noticed a substantial mass moving in from the far reaches of the solar system. As he read over the data he assumed the object was an comet. But it was moving much too fast to be of natural origin. He waved his hand to request more info. There were actually a number of different objects. These could only be ships. But...the colony was on the outskirts of civilization, one of the farthest worlds ever settled. Too far away for that fleet to be of human origin, if humans even had that many vessels.

He snapped his fingers to send an alert to his supervisor. Within seconds, a life-sized 3-D of Leela Anders appeared. She was tall, slim and thirtyish, her dark hair pulled back in a ponytail. "I'm in a meeting," she stated. "What's going on?"

"Sorry, Ma'am, but I think you'll want to see this." He waved his hand to send the readouts to her. She scanned them for a second, her eyes widening. The color drained from her face.

"Can this possibly be right?" she asked.

He slowly nodded. "I couldn't believe it either so first I ran some diagnostics. There's nothing wrong with the system or any of the bots."

The chief scientist licked her lips. "I'm sorry, everyone," she said to a half-dozen 3-D's of heads and shoulders attending the virtual meeting. "Something has come up. I need to go." Within moments the various images had disappeared. She turned her attention back to Edd. "We need to call the governor," she told him.

Within moments Leela and Edd were speaking with the governor's admin. A few seconds later a 3-D filled the wall of her office. The scene was planetside and showed a living room with plush carpeting, A man with curly brown hair sat on the floor, surrounded by several children's toys. A few feet away sat a girl, a toddler with white blond, curly hair and the same color eyes as the colony's governor, Beeja Thannel, the highest ranking human in the star system.

"Governor, I apologize for calling you so early," Anders began.

"No problem. Ginna said this is important."

"Important is an understatement," Anders said.

3

Final Hours

A FTER A QUICK DISCUSSION Governor Thannel asked the astronomers to enable him to broadcast a message to the advancing fleet. It was a non-threatening greeting in which he welcomed them to the solar system, naming himself as human civilization's leading representative in the area. He politely asked the nature of their business and if there were any way he could assist them. Because the humans were unfamiliar with the aliens' language, the scientists had their system translate it into the universal language of mathematics prior to transmitting.

The governor and the scientists waited. Hours ticked by but there was no response. The lunar base continued to track the invaders' advance through the solar system. Anders ordered the observatory's two other astronomers to come on duty.

"Ms. Anders! I think you'll want to see this," called one of the scientists.

She jogged over to his monitor. He waved his hand to instantly replay a visual that showed the fleet passing near two large asteroids a few hundred million miles away. Several of the ships fired on one of the big space rocks, vaporizing it. Moments later they took out the other. Her lips pursed in a silent whistle. She called the Thannel and sent him an e-clip of the destruction.

The governor broadcast a May Day with the details on the navy's imminent arrival and arranged for any new data from the bots and planetside to serve as a continual feed into deep space. The nearest help was a few years away by the fastest ship but at least this would provide a public record of whatever transpired.

Then he spoke another message to be beamed to the aliens' fleet: "This is Gov. Thannel again. We have established this colony for peaceful reasons. We mean you no harm. We respectfully ask that you allow us to continue to live in peace."

He strode down the hall to his home office and went about the day's tasks as best he could, but the knot in his stomach never left him.

"Sir, they're blown up another asteroid!" The alert from the lunar base came complete with a visual the scopes had picked up.

The governor sighed. He opened the sliding glass door and stepped out onto the patio. His house was on a hill looking out over the colony. The living area was a town within a pressurized dome that covered hundreds of acres. He surveyed the homes, the roads, the farms. He spotted children in a playground at the bottom of the hill. Others frolicked at a park in the distance. A small knot of people were splashing in a pool. A couple streets away, Jac Smyth was jogging with his pet, which looked like a white ball of fur. The official continued to record everything he saw and transmit those sights and sounds.

Out beyond the dome, small knots of people in *warmsuits* and respirators were scattered here and there across the orange-brown landscape, engaged in such tasks as tending to the native thornwood trees and other scrubby local vegetation or growing food in hydroponic greenhouses. Scores and in some cases hundreds of miles from the dome, other people and androids were mining mineral resources and refining them into metal. Geologists took core samples and other types of scientists conducted experiments suited to their respective fields. Hydrologists were mining for underground water and maintaining pumping stations to supply the colony. A group of geologists in a four-wheeled rover bounced along over some rough terrain.

Back inside the dome, the governor made hurried *instacalls* to the town council members. He sent a message to the entire community, informing all the workers and explorers in the outback that an unnamed emergency had arisen. He ordered them back into the dome at once. Small groups of hover vehicles and individuals using anti grav belts soon began to converge of the dome from various directions. It took some time to gather the colony's scattered workforce back to home base. The governor took a deep breath once he learned the retinal scanners had identified and accounted for everyone. Conditions outside the dome were hostile enough that any stragglers who survived the expected alien attack would eventually die of exposure.

As he looked at his little daughter, Ami still playing on the floor, a tear rolled down his face. His wife, Marva, a field scientist burst through the door. They kissed and held onto one another.

After spending several minutes comforting his wife, he sent another message into space. This one was coded.

"Roger, governor," came a bass voice belonging to Bernard Eddleston, captain of the *Pioneer,* the ship that orbited the planet. The vessel had brought the colonists to the world a decade earlier. Thannel had ordered that the star skimmer be kept in running order despite all the priorities of founding and running the colony. The captain had kept a skeleton crew to help him, rotating his personnel back to the planet surface every few months. The governor had hoped to someday begin a round of trading and cultural trips to other colonies in the immediate interstellar neighborhood.

Thannel told the captain about the aliens. The spaceman grimaced "We've been hearing chatter coming from the some of you dirt dwellers," he said. "So it's true?"

The governor nodded. "Our only hope is your ship," he said. "You're able to hide from our . . . visitors in plain sight."

"We can orbit over to the far side of the planet and cloak the ship," the other man confirmed. The cloak would render the vessel invisible and undetectable using any known technology.

"How soon can you get ready for deep space?"

The captain paused and wrinkled his brow. "A couple hours, minimum. So . . . are you thinking of escaping to another star system?"

"That may be our best move, assuming our neighbors aren't also targets. If the aliens destroy our colony the site may be toxic for quite some time."

"But . . . the colony has more people than when we arrived," he said. "We don't have enough room for everyone."

"It would be a tight squeeze but we've got to try," said the governor.

The captain continued to look puzzled. "But . . . how we will get everyone on board?"

Thannel's jaw dropped. "You're kidding, right? Why, you'll come planetside, the same way you did when we landed here and unloaded all the people and equipment years ago."

Eddleston shook his head. "That would add too much to our time line. We'd never have enough time get the ship ready, get everyone on board then get back into space before . . ." His voice trailed off.

The other man was silent for a moment. "The only alternative is to use the shuttles. But that way we could only save a few people."

"Which would be better than none. But our first priority needs to be the rest of my crew. We'll need them in order to pull this off. I'll start zinging them right away. Two shuttle craft should hold them all."

"That will leave three more for civilians. Let's get to work. Keep me informed on my priority channel."

"Yes, sir."

The governor ended the call and paused a moment. His stomach knotted in pain. He hated that so few people would get the chance to live but as Eddleston had said, it was the only way that at least a few could survive.

4

Evacuation

"OUR LIVES ARE EXPENDABLE. We'll send our shuttle planetside," Leela was shouting at the 3-D of Governor Thannel.

The governor shook his head. "I appreciate your willingness to sacrifice yourselves but there's no time. The shuttle can get from you to the ship quicker than it can get to us then back out into space."

"But there are so few life craft. You won't be able to save a fraction of the planet dwellers."

"True. But you can save six of you. Eight or more if you squeeze in tight."

"Sir, I don't feel right . . ."

"Do it. That's an order."

Her lower lip began to tremble. "Y-yes, sir. Thank you." The 3-D of the governor faded out.

Fifteen people lived on the base, so about half of them would be left behind when the end came. Leela did a probability draw to determine who would catch a ride on the shuttle. Some of the winners refused to accept their prize. She sternly insisted they go. Edd was one of the winners. She had tilted the odds his favor. He had a fiancé on the spaceship crew.

He strolled over to her desk. She got up from her hover seat, crossed in front of the desk. "It's been great working with you, Brawnley," she said, hugging him.

"Likewise, Ma'am."

"And Edd."

"Yes, Ms. Anders."

"Don't let people forget what happened here. These…these creatures can't get away with this."

"No, Ma'am." He slipped into his space suit and placed the helmet over his head then pulled a pack onto his shoulders. He turned and looked back at her, his gloved hand sticking a thumb into the air. She repeated the gesture. His head down, he trudged through the airlock and across the airless, powdery landscape toward the waiting shuttle.

Hours later, The fleet arrived at its destination. It shot to pieces the enormous solar energy collectors, the vast sails made of metal foil that were magnetically anchored hundreds of thousands of miles from the planet and beamed energy back to the colony.

Next, the aliens moved in closer to the moon. Leela was standing at the wall-sized 3-D monitor, arms crossed as she watched the sea of ships rapidly approach from afar then hover just miles above the colony.

One of the other astronomers, Vern Andresen, had several lighted data screens and spreadsheets pulled up. He moved his eyes from one screen to another and waved his hands to interact with the system.

His boss strode over to him. "What are you doing?" she cried. "We're about to die and here you are working."

He didn't even turn to look at her. "I'm preparing a counter attack."

"What are you talking about? We have no weapons."

"Ah, but we do. We don't have a way to stop the fleet but over time we can do them some damage."

"How?"

He took a moment to explain his strategy. Leela nodded. "Might work. Worth a try," The boss summoned the few remaining base residents to join in the task. Leela jumped in, too. The crew acted as quickly as possible with their precious remaining time.

After working furiously for a short time, she waved her fist at the intruders and shouted: "Do your worst! You haven't heard the last of us!" That was her final thought.

5

When Fire Fell from the Sky

BACK ON THE PLANET Gov. Thannel, breathing heavily, looked out at the sea of angry citizens. People of all ages wore warmsuits and respirators as they stood on the tarmac of the space field from which shuttles took off and landed. A dozen people were shouting at him all at once through their respirator mics. The official climbed atop a ground car and stood on the roof. "May I have your attention, please. May I have your attention." he shouted.

The uproar continued. "Quiet!" he bellowed.

The two police officers who flanked him unholstered their stunners. The crowd settled down. He had kept plans for the mini-evac under wraps and had conducted a random lottery. He had personally notified the few individuals and families invited to flee the planet but word had leaked out and resulted in this mob scene.

"People, listen to me!" he cried. "We only have five shuttles. They've already launched . . ."

"So, governor, you're going to leave this rest of us here to die?" shouted a burly man.

"Yeah, we're gonna die!" said a thin lady. Several other people shouted.

Thannel held up his hands for silence. "Not all of you. The shuttles are expected back in about fifteen minutes. They'll have time to make one more trip before . . . the invaders arrive. I'm sorry we can't save all of us. We hope to save a few. We need to let women and children go first," he said, spotting two obviously pregnant ladies in the crowd.

More shouts arose from the audience. Two police hover vehicles decelerated to a halt. Reinforcements hopped out of the cars and moved toward the crowd. Moments later, one of the shuttles touched down. The mob rushed the life raft. Thirty yards away the second touched down. A crowd swarmed it as well. Far too many people crammed their way into each of the boats. The first one lifted off, followed by the second. One by one, the other lifelines into space touched down. Once all of the vehicles had filled to triple capacity and lifted back off, many agitated people were still left standing on the ground. A woman with dark, straight hair and a swollen abdomen had been left behind. Some people shouted at the departing shuttles. Others picked up rocks and threw them but the vessels were already high in the air and rapidly shrinking. The mob began to dissipate. Thannel shook his head, jumped back onto to his air speedster and streaked off toward the dome.

Within minutes of his arriving back at his house, friends and neighbors began to arrive. Some talked animatedly but others were silent and somber-faced. Many hugged one another. A number of them were weeping. The crowd in the governor's living room became so thick he led everyone outside and into the large yard. He tried to call for silence but no words would come.

Finally, he was able to say: "Thank you all for coming. Not quite ten years ago we landed here to found this colony. I'm proud of what we've accomplished. I'm proud of all of you, our neighbors and friends. My wife, Marva, my buddy Jac, it's been great knowing and working with all of you. I know you're aware of the imminent threat to our lives. We've done everything we can to communicate with these beings and they've rejected our requests for peace. We've been able to evacuate a few people but due to time constraints we could not save everyone. I've ordered that the strongest possible force field bubble be set around our colony. The rest is up to God. If you'll all join hands I'd like to lead in prayer."

T-minus fifteen minutes.

Ten minutes.

Five.

One.

6

Like Ripples across Space

N EWS OF THE UTTER destruction of the Rantran colony gradually spread among humanity like a stone hitting a pond and creating ripples. Word spread at the fastest known speed, the speed of light, but still took years to reach across interstellar geography. A few years after the attack, the *Pioneer* arrived at a nearby colony. The survivors gave eyewitness accounts of the alien fleet and their brazen scorched earth attack.

Meanwhile, other local star systems began to receive Rantran's May Day and the 3-D video of the incident. Some local star systems, as they were able sent drones or manned ships to investigate. They confirmed the survivors' story: the attack seemed unprovoked and the devastation was total. Over time, several starship crews who had seen the aftermath sent transmissions, adding to the growing accounts of the atrocity that became known as Black Death Day.

Fear led the gradual expansion of humanity among the stars to come to a halt as plans to colonize new star systems were put on hold or abandoned. Interstellar navies added more ships and began patrols along the vast rim of human habitation. The military began the construction of various deep space star bases and armed stations. Defense budgets on numerous planets increased sharply. The tension in some of the outlying sectors became so great that trivial differences and simple mis-communications led to skirmishes between normally friendly star systems.

Most of humanity lived within a fifty light year radius of the cradle of civilization, a blue-white planet known as Earth. Within that fifty light year span were approximately 1,400 star systems that included about 2,000 stars. Outside of the Sol-like stars, most of the others were red dwarfs.

The most heavily inhabited systems were the 135 or so that included stars similar to Earth's sun, and many of them hosted Earth-like planets (ELPs). In each inhabited system any ELP typically became the heaviest-settled, followed distantly by any Mars-like planets (Mlps) then some of the major moons. Rantran III was a pebble of a world that orbited a single yellow star. It was the only planet in that solar system that was semi-hospitable to human life. It wasn't an ELP and it wasn't an MLP. It was somewhere in between, although the art and science of terraforming had been gradually giving it more Earth-like characteristics.

During the two millennia that humans had been colonizing the stars, Man had come into contact with a mere handful of other intelligent species. Numerous planets contained abundant varieties of life but no sentient beings. This had proven true of many ELP's. Had God designed them to accommodate the eventual expansion of humans into space? Most non-human forms of intelligence had arisen on planets that that were too sweltering, two frigid or too radioactive to be easily habitable by humans. But most of these non-human races were too vastly different than humans for the two groups to easily communicate with or relate to one another. And non-human civilizations tended to be either so advanced as to be virtually incomprehensible or else very primitive. Thus, no non-human culture had ever been found to complement humanity. Or threaten it. *Until now. Was the attack an isolated incident or would there be more to come? If so, when? And where?* As the media dwelt continually on the local genocide and its ramifications, the public on many worlds demanded action. *Yet, how could the officials react effectively when virtually nothing was known about the attackers?*

Black Death Day was the day human history changed forever. This is the story of the tiny band of men and women who took the first human action against of the murderers of more than three thousand colonists who perished that day in an isolated corner of space.

7

Light Years from Love

"WAS SHE TRYING TO seduce me?" Captain Erik Houston wondered. About a month earlier, his ship, *The Initiative,* had departed its last port of call, the planet A'laama. During the mission to that world the captain and the female head of the planet had been stranded overnight in the wilderness. The two of them had slept apart, he is the ATV and she in a small tent. Nothing physical had happened between them. But by the time the mission to the planet had ended, their feelings for one another had grown and it had been difficult to for the spaceman leave.

Now, weeks into the ship's return trip back to base, life aboard ship had slowed down and he had decided to stay busy by cleaning the interiors of the ship's vehicles. It was a task he would normally assign to a crew member, but something kept telling *him* to do it. He walked through the air lock into Landing Craft Number Three, still containing its pressurized atmosphere. Within the large craft sat the ATV he had driven months earlier. He smiled at the more pleasant of the memories from that crazy trip. He opened the passenger hatch.

He did a double-take when he saw *it.* That mysterious black overnight bag she'd brought with her. She'd been reluctant to tell him of its contents. He didn't think she'd taken anything out of it the entire trip. He placed the item into a large trash bag and went about the cleaning chores. Unable to concentrate, he grabbed the bag, hurried out of the vehicle and left the air lock, heading back to the main living quarters. Keeping an eye out for wandering crew members, he casually walked down the hall, holding the trash bag as he approached his cabin. He looked both ways down the hall before opening the door. He stepped inside, closed the door and leaned against it, exhaling with relief.

He opened the sack and pulled out the luggage, placing it on the floor. He sat down on his hover bed and stared at the item. He laid back and tried to go to sleep. He opened one eye. He felt the bag was watching him. He rolled on his side. It still haunted him. He grabbed the black container and placed it on the bunk. He ran his thumbnail down the bag's seal strip and unfolded it open. The scent of her favorite perfume, Jynspice, wafted to his nostrils, bringing back other memories.

Setting atop the contents was her electronic organizer. A square with a given date was enlarged.

It was the day they had driven into the wilderness. A little familiar with A'laaman devices, he pushed lightly on the square. A mech voice ran down her list of scheduled activities for that day. Some of them sounded fairly important. In addition to her arrangements with him, she'd had five appointments. There was even a warning message in another voice saying that the trip with Captain Houston conflicted with the other events. She had overridden the message and noted that she had planned to keep the meeting with the spaceman.

Puzzled, he checked the dates two or three days on either side of their drive into the jungle. She had scheduled fewer events for each of those days, yet she had chosen her most committed day to be with him. Scratching his head, he pulled the original date back up. It also contained a note that exceptional, once-in-several-year sunspot activity was likely that day. And could potentially disrupt communications for one to two days. The prediction of bad space weather had come true, knocking out the GPS the captain had been relying on for navigation, leaving their ATV lost in the wilderness. She had purposely chosen the day that offered the biggest chance of being stranded in the outback with him!

Trembling, he shut the bag. He took a deep breath and opened it back up. He removed each item and laid it on his bunk: a hairbrush for her lustrous mane, a tube of sun minimizer, a makeup kit, a couple spare V-screens (the natives' hand-held communicators), two neatly folded changes of clothing, some toiletries. He reached into the bag and pulled out the next item, a self-chilling bottle of sparkling intoxicant. As he picked it up, it began to cool and within moments condensation was forming on the bottle. He set it back down. At the bottom of the bag, wrapped in layers of protective cloth were two crystal drinking glasses engraved with the A'laaman letter Z, which looked like the standard letter Z but with a diagonal line through it.

He burned from his head to his feet. He started having graphic thoughts about her but pushed them out of his mind. He shook his head.

They had never done anything physical beyond a few hugs. Plus a couple passionate kisses his last night on the planet. Had she wanted to do so much more that long ago? He began to grieve over the situation. Was she not what she had seemed? Was she a hypocrite? She was a passionate woman who had appeared to trust in God. Erik had even prayed with her. He still felt she was a good person and was reluctant to judge her.

But surely he had made a mistake leaving her behind on the planet. He stood up and for a long moment thought about breaking all interstellar protocol and ordering his flight director to do a one-eighty. But the ship was more than two light years out. It would be impossible to get his bosses to understand his reversing direction. It would end his career.

He returned the contents to the bag, sealed it and placed it in a secret compartment, setting the electron lock. His next stop was the ship's workout room. It was going to take him a long time to unwind.

After a vigorous workout and shower, he dragged back to his quarters, his muscles protesting all the way. He waved his hand to set the hover field's comfort level to forty-five then eased into a lying position. His eyes facing the cabin's low ceiling, he began to wonder if there was some way he could ever get back to her planet.

He had served for years as commander of a First Contact ship the Association dispatched to visit star colonies. Before he had left for his last mission, the brass had implied he was up for a promotion. That typically meant heading a Followup ship. When he delivered his report on his last mission, he was going to recommend the world as a strong candidate for immediate Followup. But the bureaucracy moved slowly and there were no doubt other planets already in the queue ahead of A'laama. The fastest he had ever known the high hats to take action on a Followup request was a decade. He knew that by the time his crew and he arrived for shore leave a few weeks hence, five years would already have passed back on her planet but only two months for his crew and he given the *temporal dilation* caused by their travel at almost the speed of light. Ten years, best case scenario, for the bosses to dispatch a Followup ship and an extreme best case scenario would have him heading up the vessel. Then five more years of planetary time to get there. So, by the time he saw her again she would no longer be in her mid-forties, a similar age to his, but her mid-*sixties*. And if he got back there, would she still be unattached? Would they still relate to one another?

Too bad he wasn't interested in her daughter Omma, a vivacious college student with long, auburn hair. She had seemed to have a crush on him when he was on A'laama. By then, *she* would be about the right age. What if . . . He clenched his fist, ashamed of having those types of thoughts.

The captain rose to his feet and went into the hallway. He strolled until he finally came to a port. He gazed out at the stars. Staying behind on the planet had not been an option. Turning the ship around was not an option. He would never see Zama again, at least not the way he remembered her.

$\mathcal{8}$

Suicide Mission

SHORE LEAVE BEGAN NORMALLY enough. More than a decade had passed planetside during the ten light year round trip that had encompassed the previous mission. But the captain and crew had only aged a total of nine months, including two months shipboard on the trip to the planet, a five month stay on that world then the two month return trip to regional headquarters, all due to time moving much slower than "normal" on a ship traveling at close to the speed of light. Returning planetside after all those years made a star jumper feel like a time traveler. Erik was always fascinated with catching up on years' worth of news that had taken place in his absence. But that could come later. First priority was R&R.

The day after his ship docked at its orbiting port, Erik and his second in command, Lt. Federico Montoya caught a tourist flight to a neighboring world in the same solar system. Hours later, the duo was among a handful of others riding in a transparent-walled floater, a craft that navigated the currents of the upper reaches of the planet's turbulent atmosphere. The immense pressure would instantly crush the fragile vehicle if not for the force field surrounding it. The floater bounced and heaved, the passengers all held in more-or-less place by ener fields. Erik laughed as the vehicle jerked them one direction then lurched in another. His stomach plummeted as the floater dropped a thousand feet within seconds. The craft's unseen crew kept a fine balance between thrilling the passengers and allowing too many of them to get motion sick. Fred already looked under the weather despite the meds he had taken.

In front of them and on all sides was the murky, swirling atmosphere. Lightning from distant storms occasionally illuminated patches of sky. A bolt flashed in front of them, sending shivers up Erik's spine. He elbowed Montoya in the arm and pointed. In the distance was one of the native creatures. Somewhat resembling an enormous bird with over-sized wings, the being navigated the currents and eddies, sometimes rising rapidly and falling just as quickly. Montoya gestured toward two others, far-away dots.

Erik waved his hand and a closeup of the nearest creature appeared at eye level. Its head contained two beady eyes, two long, extended feelers and a vice-like mouth. A smaller animal, resembling a bat with elongated wings, fluttered within twenty yards of the larger animal. It whipped a feeler out to its full length, snagging the hapless flyer and depositing it into the mouth-vice.

Lightning struck one of the more distant birds, instantly frying it. Several of the passengers applauded while others whistled and hooted. Seconds later, shock waves jostled the floater.

Upon recovering, Erik thought how much fun it would be to hunt the giant birds but he realized they were a protected species. Fred and he planned both holo and live hunts of other animals later in the week.

Erik jumped as a verbal page blasted out his name: "Houston! Urgent message for Starship Captain Erik Houston!" The captain set up a sound block. The message was coming in coded. He set up a descrambler.

"Houston here," he growled.

A 3-D of a face appeared in front of him. He recognized it as the regional military director for this sector, his boss' boss' boss. Houston's stomach clenched. The man must be calling from the system's main planet, the one where their ship had docked at the orbiting maintenance station. It would have taken the light and sound from the message a couple hours to travel this far. Since it would take as long for Houston's response to reach the location, this was likely a one-way comm. Still, why was local HQ calling? Protocol demanded a week after debrief before a captain reported in further. You don't interrupt a man's leave. That just wasn't done.

"Houston," the officer bellowed, "You did an outstanding job on your last mission. Congratulations, son. We've had our eyes on you for a long time."

The captain balled his fists. They were calling from billions of miles away just to say "Good job?"

"In fact," the gravelly voice continued, "you've been so effective that's it's time for you to move up. We have an exciting new command for you. A vitally important one. I want to see you in my office in 0500 hours. Gather as many of your crew as possible in the next day or so and make sure they stay close by. That is all," he concluded, saluting.

"Aye, sir," he replied, returning the gesture, knowing that by the time the superior saw it he'd be well on his way back to the planet. The 3-D vanished.

The captain waved his hand to end the sound block. He sat with red face and folded arms.

"Trouble?" asked the chestnut-haired Montoya, wrinkling his brow.

"The brass has ordered me back to HQ," muttered the captain.

"What's up?"

"Don't know, but don't like the sound of it."

"I'll go with you."

Erik nodded. Montoya slapped him on the shoulder. The captain placed a 3-D call to the nearest inter-planetary transport firm.

Three hours forty minutes later, the return ship was piercing the dominant planet's atmosphere. As Montoya stared straight ahead, the captain gazed out the port, emotionally numb. Once the transport had descended to within twenty miles of the surface, he spied the lighted towers of the premier city, looking in the distance like a miniature, enchanted kingdom. The towers all seemed to be made of light. As the ship continued decelerating, the true scale of the towers became evident. The city was called Newhattan. Centuries earlier it had gone by New Manhattan, having been named for an Ancient Earth metropolis because of the predominance of soaring towers. Locals had eventually shortened the official name to reflect the common usage.

Once the transport landed, Montoya headed off to seek some entertainment while Erik took an automated *hover cab* to the coordinates of Star Fleet's regional headquarters. As the cab took off, a twenty-foot-tall headline scrolled across the sky: *Military Appropriations For Next Year to Increase Twenty-Five Percent*. A second banner followed: *Navy Developing New Weapons to Address Alien Menace*. Inside the vehicle, a 3-D newscast was blaring about the latest public opinion polls tracking a high state of alarm among the populace. He tuned it all out.

Moments later, the cab touched down in front of a massive building. The structure had no obvious entrance, no apparent windows. He walked up the facade and held up his palm. A camouflaged door swung open. He stepped inside.

"Welcome, Captain. We've been expecting you," said the well-modulated voice of a military android. Houston had never in his career reported to a facility with such tight security. In a narrow, dimly-lit hallway, an enormous officer ordered him to pass through a full body scan, a retinal scan, a second body scan and a rather intrusive mind probe that he could actually feel inside his brain even though the probe did not utilize any physical devices. A med scan followed all of the others. Houston was certain there were other remote probes and scans taking place. The hallway ended in a solid wall. The wall split into two halves that parted to create a doorway. Two officers wearing immaculate uniforms and side arms stepped into the hall. They motioned for the captain to follow them then they stepped back through the doorway. He followed them into the next area. The two halves of the wall merged back into one, soundlessly and seamlessly. The escorts led him through a maze of hallways. They stopped in front of a massive door made of inch-thick glass.

"Sir, Admiral Lefkowicz will see you now!" said one of the officers. He pointed at the door.

Erik saluted them then pushed open the door. A stocky man with a grey crew cut and a square jaw glared up at him. It was the man who had sent him the video call just hours earlier. The admiral had grown up on a planet of the Gliese 581 system, a world that was seven times more massive than Earth. The man's somewhat squat, compact appearance was typical of someone who had grown up in a high grav environment.

Houston was typically gone eight to ten Standard planetary years per mission, a much shorter time for him due to the time dilation. He had first become aware of Lefkowicz three or four missions ago when the future admiral had gotten his first senior position with the regional brass. Over the decades, he had continued to move up in rank.

"Sir, Captain Erik Houston reporting as ordered." he snapped, saluting the high commander.

He returned the salute. "At ease, Houston," he growled. "Have a seat."

Erik complied.

"First, we'll have a briefing." The admiral waved his hand and a 3-D video appeared. It showed a small town on another planet. The view panned to show a number of adults and families going about their daily activities. Abruptly, the idyllic scene ended with shots of fire falling from the sky, followed by scenes of blackened earth, smoke-filled skies and total devastation.

"Sir, who committed this outrage?" Houston cried.

The officer shot him a penetrating gaze. "Non-humans," he replied. "A massive fleet was passing near the colony's solar system when it launched the attack. Our settlement appeared to be a target of convenience. Or maybe they viewed the loss of life as some twisted sort of fun," the gravelly voiced man spat.

Houston shook his head.

Lefkowicz waved his hand again. A huge sphere filled the room, a schematic of interstellar space for dozens of light years around. Near the upper left hand edge of the sphere was a mass of illuminated yellow dots. Some distance away was a similar sized group of tiny blue lights. "These are two navies, each one belonging to a different alien race. The one on the starboard side destroyed our colony. We've been tracking communications between these two forces. We've used our most sophisticated de-scramblers and these two species appear to be taunting one another, challenging one another to war back at the one race's home planet. The beings that destroyed our colony."

Houston's jaw was hanging open. "Sir . . . where do *we* come in?" he asked.

The officer leaned into Houston and explained his new assignment. A fire grew in Erik's belly. "Sir . . . with all due respect since these two species are about to go to war, why not wait first to see what damage they inflict on one another?"

The fire in the admiral's eyes was even more intense than the one Houston felt internally. "Because these aliens" (he spat the world) "have humans on dozens of worlds scared of their own shadows. The public is demanding we do something and our political leaders are squawking even louder. The soonest we can get a ship, *your* ship on the scene is about the time those two races will start shooting at one another, but in the confusion it will be the best possible time for you to do what you need to do."

"Or, the aliens will be on such a heightened state of alert that the danger to us will be exponentially higher," Erik thought. His stomach knotted until it hurt. He was tempted to decline the job even if it meant resigning the Navy. *"We won't come back alive,"* he thought, his fists clenching. He wanted to serve humanity but was reluctant to put his crew in that level of danger.

"Sir, permission to comment."

"Granted."

"Sir, with respect, I'm not a *Special Opps* guy," he said as his orbs met the other man's steely eyes.

"Nonsense," the boss replied. "You've conducted successful military opps over the years and have hardly ever lost a crew member. Look at the last planet you were on. What's it called? A'laama?"

Erik's head dropped. "Yes," He returned to piercing the other man's gaze. "A'laama. They'd make a prime candidate for a second visit. I was hoping . . ."

"All Followup trips have been canceled due to the Emergency," the big man snapped. "We're putting most of our resources into defense just in case. So, you have your assignment. You're being relieved of your command of the *Initiative*." (Erik swallowed hard when he heard this.) "We're giving you a new, state-of-the-art ship. You and your crew will receive some brief but intense training on the starbuster and its offensive and defensive capabilities. Your mission launches in 9600 hours. That is all."

The captain felt as if he had been kicked in the stomach. He stood up from his chair and saluted. The commander returned the gesture. Houston turned to leave.

"God be with you, son," the admiral called after him.

The other man stopped in mid-motion. Electricity ran up his spine. "Thank you, sir." he said quietly.

After following a tall android down another maze of hallways, he stepped out into the bright sunlight, shielding his eyes with his hand. He was oblivious to the bustle of the city. If his mind had not been occupied with other matters, he would have noticed that several of the most prominent buildings hadn't existed the last time he'd been here. His shoes began to pound the sidewalk. After walking about twenty minutes, he became aware of his surroundings. A few blocks ahead stood a tower that seemed to be made of light, typical of a modern skyscraper. The top of the structure disappeared into the clouds. He strode over to the building and stepped into one of the *lift shafts*. He began to ascend. As he accelerated, he shot past dozens, hundreds of floors but without any sensation of motion. Huge numbers on either side of the shaft noted his progress. The shaft was moving him so fast the numbers were barely discernible. Realizing that the top floor was approaching, with split-second timing he hopped out of the shaft. His feet hit the solid floor so quickly the impact caused a sharp twinge in his right knee. His gait slowed a little as he headed down a plush carpeted hallway while trying to work out the kink in his leg.

A *suicide mission*. That wasn't what the brass had called it. They'd given it nicer, politically correct terms. His career had consisted of one-shot visits as head of a First Contact ship, following up on colonies founded generations, even centuries earlier and kept isolated by the vastness of space. He had visited dozens of different worlds, no two of them the same. Some missions had gotten his crew and he into life and death circumstances. But never had he been ordered on into certain death. And this was what the brass called a promotion!

The big boss had been quite plain. No married couples on this ship. Singles only. The brass had offered a fortune in combat pay and life insurance to all who signed up to go. A fat lot that meant if there was no one left behind to inherit it.

Erik jiggled the drink glass, causing the remnants of the ice cubes to race around in little circles. He'd been staring at the amber liquid long enough. He tossed down the drink, feeling the burning down his throat, into his stomach and out to his extremities. Bimaanian resa liqueur. The best stuff in this sector. Maybe anywhere.

The spaceman looked up from his drink and gazed out the glass-like walls. He was on the 695th floor of the Triumph Tower, the tallest monument to man on the planet. Outside the transparent walls a bank of clouds, looking like a fluffy carpet, filled the horizon. Peeking through this false floor was a scattering of other buildings. Some were miles distant but they still looked close enough to touch. It was late in the day and the sun was just beginning to turn a portion of the cloud floor a lemon hue.

He pictured *her* face as if she were sitting across from him. A face that could be as studied as that of an ancient poker player. Round, slight double chin, a bump on her nose from a gang attack as a teen. Brown eyes that could be steely with determination, laughing mischievously, or wide with curiosity. A neck that was always adorned with a gold cross necklace. Soft, light brown hair flowing past her shoulders.

He thought of how aloof she had seemed when they had first met. He thought of her fiery temperament and how she'd clashed with him on several occasions. He had finally fallen under her spell. But it had been a match that couldn't last. The ruler of a planet and the leader of a starship. Two different lives with no way to integrate them.

It had been two months since he'd left her behind. Two months for him, five years for her.

Curse Einsteinian Relativity! She'd had a long time for life to intervene. *Did she still feel the same about him?*

The captain shook his head while the bar android chortled in amusement. The star man returned his gaze to the transparent walls and sighed. The cloudy carpet had become golden with a little pink at the edges. His heavy heart was unable to appreciate the beauty that was before him. Fighting back a tear, he continued thinking of the one person with whom he'd like to share this sight. Light years away and inaccessible. The woman it was impossible that he'd ever see again. Even if he lived.

A few hours later Erik was pacing the room, his eyes viewing the seated crew members who had signed in for the virtual meeting, sending 3-D images of themselves to the captain's location. The meeting would consist of multi-way conference.

The captain was slightly tall and had powerful shoulders and a narrow waist. He had dark hair and was clean-shaven but was prone to five o'clock shadow, sometimes giving his face a greyish appearance. His somber mood added to the impression.

His eyes rested on several crew members. Luci Strong was short, blond and thirty-ish with a round, pink face. She worked out frequently and had close to zero body fat. Her face reflected a serene beauty. She'd been on the crew for years and was one of his most capable officers. She haled from a star system where the inhabitants' accent sounded like an old Southern drawl from ancient America on Earth. After all her years in space and visits to dozens of planets, Luci still had the accent.

Seated next to her was her best friend, Marji Faubner. She was medium height and had light brown hair that she sometimes wore in a ponytail. Luci and Marji were like sisters but without the fighting. Marji had a sharp mind and had been one of Erik's senior engineers aboard the *Initiative*. His only engineer who had ranked higher was Daj Minj, who had fallen in love with an A'laaman woman and had stayed behind on the planet. Erik paused a moment and wondered how Minj had been doing since the *Initiative* had departed.

Erik shook his head as if to clear it. His eyes fell upon Irv Malvo who was average height, balding, and grey-headed. Over time, Irv had appeared to be ageless. He looked sixty but was past ninety. He was Erik's longtime security chief and spiritual advisor. Normally a jolly man, to-day Irv was poker-faced, seeming to sense the gravity of the situation.

Standing next to the captain was his second-in-command, Montoya, the one person with whom he had so far discussed the mission. He had chestnut hair and a Van Dyke beard, which he was stroking with thumb and forefinger, his eyes having narrowed to slits of concentration.

The leader continued to survey the crowd as more crew members signed into the meeting. He had hated to interrupt everyone's shore leave so early on. It had been over a day since the crew had disembarked and by now, they could be anywhere on the planet, perhaps orbiting it or even elsewhere the solar system. His hand-held device noted forty-eight of the fifty-two were virtually present. Close enough.

"Thank you all for coming," he started off, louder than he had intended. "I apologize for the short notice," he continued, clasping his hands behind him as continued to pace. "Ladies and gentlemen, the brass has relieved me of my command of the *Initiative*."

A collective gasp echoed around the room. The captain had been in charge of the vessel for more than a dozen years as measured by the odd combination of shipboard time and the collective time spent on various planets. He held up his hands for silence. "I've been offered . . . actually, *ordered* to assume command of a newer, sleeker ship. But the nature of my orders have drastically changed. I've been transferred to Special Opps. Our assignment will be far more dangerous than any we've previously undertaken: re-con on an alien planet where two rival species are expected to go to war." He looked around the room again. The silence was palpable. A number of mouths were hanging open. "You're probably wondering why we humans are getting involved. One of the alien species has struck us hard." He waved his hand to show the 3-D of the destruction of the Rantran colony. The crew sat in stunned silence as the record of the brutal attack unfolded.

"I won't mince words," he continued after the video ended. "There's a better than average chance we won't come back. Our new ship has a crew capacity of only sixteen. I appreciate all of your hard work and faithfulness over the years. We've been like family. I don't feel right *ordering* any of you to take part in this mission. So I'm asking for volunteers. We launch in less than four days. The rest of you will be assigned to other ships."

Montoya, who was already standing, immediately held up his hand. Luci sprung to her feet, as did Irv and several other men. Marji immediately followed Luci. Within a moment, more than half the attendees were on their feet. Three-quarters. Practically everyone. The captain swallowed hard, overwhelmed by the display of loyalty. The group broke into applause. A few cheered.

Houston lowered his head, shaking it from side to side. He again raised his hands for order. "Thank you," he muttered. Then, slightly louder: "I appreciate so many volunteers. Lt. Montoya and I will make the final selection. You'll all be notified as soon as possible. That is all."

Over the next couple hours, Erik and his first officer discussed the qualifications and loyalty of each of the volunteers. After a detailed and sometimes animated discussion, they came to an agreement on the crew. In addition to the captain and Montoya, officers would include Irv Malvo in charge of logistics and Marji Faubner as chief engineer. Marji would also serve as flight director supplemented by two backup pilots and the ship's autopilot. The versatile Luci Strong would be the head of communications, assistant head of logistics and, when needed, helping as a med tech. Luci typically fulfilled several roles on a given mission and had been exemplary at all of them. Rounding out the crew would be several enlisted men and women for a total contingent of sixteen. The leader hated to dismiss more than two-thirds of his old team but was grateful that whatever their re-assignments, they would likely entail less risk than what the sixteen would face.

9

Phantom Lady from the Past

THAT EVENING, THE ENTIRE crew was physically present in a hotel suite for a goodbye party. At one end of the room was a levitating, rotating cake shaped like the *Initiative.* The mood in the room was palpable when the hotel staff began to slice up the dessert. A hover table contained a punch fountain and some light snacks. Several knots of crew members talked animatedly. Some of the men and ladies were dancing to lively music. Others were in a glass-walled *zero grav* room, laughing and floating various objects around as they, also floated from one part of the weightless zone to another.

Marji was slow dancing with Jev, a tall, blond man with long sideburns. He had been her fellow crew member. They'd had an on-again-off-gain relationship for years. They had drawn close on their last assignment but Marji had learned Jev would not be going on the new mission.

In another part of the room Luci, wearing a light blue gown and a silver pendant necklace, glanced around at the festivities. *Where was the captain?* He must have ducked out shortly after the crew had done a comedic roast of him followed by android waiters serving cake and punch. Houston had seemed unable to relax even at the party.

Luci hurried out of the building and waved her hand to summon an auto-cab. By the time she reached the sidewalk, the cab was hovering a few feet above the ground. A round entry port appeared in the side and a ramp descended. She hiked up her gown at the knees and hurried up the ramp. The vehicle launched skyward without creating any sense of motion. The city of light retreated until it looked like a miniature carnival.

The cab sped to the destination Luci had pre-entered. The darkness of the night quickly gave way to the blackness of space.

A short time later, the cab touched down hundreds of miles above the planet at an orbiting station. The star port was deserted at this hour. She slipped off her high heels and held them in one hand while she jogged across an ice cold floor over to the cavernous bay where the *Initiative* would be serviced prior to being turned over to its new captain. Standing off to the side, staring at the vessel in the dark was a lone figure. He stepped up to it and stroked the hull.

She crept up to him and laid a soft hand on his shoulder. "It's been a great ship," she said.

He smiled without turning around. "The best. How'd you know I'd be here?" he asked, still looking at his former space home.

"Where else would you be?"

Silence.

"You're missing a good party," she said.

He didn't respond.

"Captain, I know you have a lot on your mind. But the Skymasters are playing in Betaville tonight . . ."

He turned and looked into her eyes. "You know my rule about dating crew members."

"It wouldn't be a date. Look, I know you're still thinking about that lady and I wouldn't mess with your mind. This will just be two friends, two coworkers looking for some R&R."

He sighed. "You win. But aren't we a little too dressy for a fly hockey game?"

"Who cares?" she laughed with a shrug.

They walked silently from the bay and found a shuttle heading planetside. They were soon zipping through the sky a few thousand feet above the planet surface. The stadium was one of several buildings that hovered far above the city. The vehicle approached the arena and descended toward a wide ring of concrete that surrounded the building and served as a landing pad and parking area. "Say, isn't this place new?" he asked. "I don't think it was here last time we were home."

His partner laughed. "No, it's over twenty years old."

"You sure?"

"Yes."

"When I come home, I always feel like we're in a time warp."

"That's because we are," she drawled.

"Game's about to start. Let's go," said Erik, glancing at his wrist chronometer. Luci and he hurried out of the vehicle and into the sports venue.

Sky hockey was modeled after the ancient ice hockey except that it was played in three dimensions by players who soared through the air on anti grav belts and used magnetic poles to guide a metal ball to a goal. Erik had always been a huge fly hockey fan. As the game progressed, Luci was glad to see him begin to relax. He rose to his feet and shouted when his team scored strategic goals. She even joined in some of the mayhem.

Hours later, the captain had a smile on his face as they headed toward an exit. She looked up at him and smiled, too. *Mission accomplished.*

He strode over toward an empty part of the lobby. "Will you help me with some shopping?" he asked.

She raised an eyebrow. *The captain? Shopping?*

He waved his hand to set up both a visual block and a sound block. A virtual jewelry store sprang up around them. He walked over to a case displaying dozens of expensive-looking rings. She looked at him even more uncertainly.

He pored over the selection. "Are there any that you like?" he asked.

"Any that *I* like? Uh . . . that one's nice. But Captain. What . . .?"

"I've got her ring size from the security scan we did of all major officials on her planet . . ." he said absently. His eyes lit up. "Whoa! Look at this." He pointed at a ring with a large, elegantly-cut local gemstone, ice blue in color. "Do think she'll like it?"

Luci was breathless. "I-I think any woman would be flattered."

"I'd like to see that one, please." The incredible ring appeared in the palm of his hand. He studied it, turning it over. "Size 11." The ring re-appeared in the display case and a slightly larger one materialized in his hand.

"How much?" asked the captain.

"Thirty-two thousand units," said a dis-embodied female voice.

"Twenty-seven five."

"Thirty-one," the unseen female said more firmly.

"Twenty-nine. You're not the only jeweler in town."

"Thirty-even."

"Twenty-nine seven-fifty."

"Done," the disembodied voice sighed.

He waved a coin-sized debit chip as payment. A decorative box appeared around the ring. He stuck it in his pocket.

Luci stared at him, slack-jawed. Her normally-sensible boss had just paid a fortune for a ring for a woman he would never see again.

10

One-Man Mutiny

Montoya was pacing up and down, his face red. *What was the captain trying to do, get them all killed?* They were days ways from leaving on a critical mission. Yet all the boss seemed to think about was that planet dweller. Montoya had never liked her. Temperamental aristocrat. He'd hoped that after the months on the return voyage the boss gotten her out of his system. A First Contact ship like their old vessel never returned to a previously-visited planet. Zama Elle was ancient history, as if she had never existed.

Plopping into a chair, the second-in-command was almost trembling. He swallowed hard as he thought about what he had to do. He hesitated for a few minutes until the fire in his stomach got best of him. He waved his hand and placed an audio-only call, saying in a low voice: "Central Command? Yes. I-I've got some critical information affecting Mission Alpha-Four-Gamma."

An hour later, he was standing in person before a long desk of dour officials. He wore a grey cloak, its hood covering his face.

"Lt. Montoya, what is this all about?" demanded the lead official.

"And it's late," snapped another. "Can't this wait until tomorrow?"

He pulled the hood off his head. "Sir, I have grave concerns about Captain Houston's ability to lead this mission. He's mentally unstable."

"On what grounds do you make this accusation?" shot the second dignitary.

"He thinks he's in love with the ruler of the last planet we visited. He kept talking about her the entire return trip."

"So, where's the mental instability?" snapped the third council member, a lady who wore her auburn hair in a bun.

Montoya pulled at his beard. "Ma'am, he just bought a ring for her. An expensive ring."

"And you saw him do this?" asked the council leader.

"No. Another officer did. Then she confided in me. She's concerned about him, too."

"Dr. Montoya," said the leader, "Captain Houston has effectively served the Association on dozens of missions spanning centuries of planetary time. That's why we chose him to lead this mission. We do an auto psych scan of each captain when they first arrive for leave. This is frankly none of your business, but your commander is well within the range of normal. Everyone has their own way of reacting to the stress of conducting a mission, particularly one like this. If the captain wants to throw away money on some phantom woman from the past, so be it!"

The lady commander jumped back in: "And might I remind you, lieutenant, your recent track record has been less than stellar. At your last port of call, you and engineer Daj Minj engaged in a minor mutiny that temporarily cost us the goodwill of that planet and almost scuttled the mission. For this you were both disciplined and demoted one full rank. Is this not correct?"

Montoya dropped his eyes to the floor. "Yes, Ma'am."

The third official jumped in: "Given your previous lack of loyalty, you should be glad we're not pulling you from the mission. Now get out of here while you still have an assignment."

"Yes, sir." He shuffled out of the meeting room.

Barely ten minutes after he left the meeting a 3-D message in tall red, letters raced in front of him just inches from his eyes: *"Montoya! My office! Now!"* He cursed before responding. He hustled over to the temporary office the Navy had assigned to the captain for the few days until the mission launched. The boss was standing and drumming his fingers on his desk when Fred burst into the room. Both officers were equally red-faced.

"Take a seat," the captain ordered.

He complied.

"What's the meaning of going behind my back to complain to the brass about my leadership?" Erik demanded. The first officer looked at the floor. "Well?"

"It's that woman," Fred cried. "That Zama woman. Erik, you *bought her a ring?*"

"Is that any of your business?"

"It is when we're about to leave on a mission. I need you, *we* need you to be thinking soundly, not day dreaming about sex or romance or *whatever* is going through your head. Humanity needs strategic information on these hostile aliens. And fifteen other people are trusting you with their lives."

"If you had concerns, why didn't you come to me?"

"Sir, I was concerned about the soundness of your judgment."

"No excuse. I'm your commanding officer."

Montoya glared at his boss then dropped his eyes. Finally, the captain resumed speaking. "Look, my mind and heart are totally on this mission. I'm determined to get it done right. I'm committed to bringing all of you back home safely."

"Or die trying," said Montoya, reading the captain's mind.

Houston ran his hand through his thick hair. "But I'm also deeply disappointed in you. First that stunt on A'laama. Now this. I can't keep having you stab me in the back. If you want reassigned, just say so."

"No, no. I want to go on this mission."

"Then act like it. I want you on the mission, too. You're definitely the most qualified. You just said you want me to have your back. But I need you to have mine, too. So, no more second guessing. You've got to trust me."

Montoya sat in silence.

"So, are you with me?" snapped the captain.

He nodded.

"Okay. But any more disloyalty and you'll force me to replace you. I can even do that mid-mission if I have to. Are we clear?"

"Aye, sir."

"Get some sleep. We've got an early date with the new ship tomorrow."

Montoya arose from his chair and shuffled out of the room.

11

Starbound

MORNING CAME ABRUPTLY WITH the first day of intense training and drilling beginning at 0400. Several of the crew members groaned due to the pre-dawn time, having stayed up late for the crew party the prior evening. Their training director was a trim man with a salt and pepper crew cut. He wore slacks and a snug black T-shirt, appearing to have even less body fat than Luci. As the training unfolded, the captain and crew became familiar with their new ship, a sleek model that could reach incrementally closer to the speed of light than even their previous vessel. The starjammer was designed to get them to their mission and back out again as quickly and safely as the latest tech allowed. Rather than using the traditional landing craft to get crew members to a planet's surface, the vessel would create a temporary lift shaft that would serve as a space elevator from the ship to the planet and back again. The vessel had some enhanced weapons systems as well. The instructions began with 3-D immersion sims followed by live training and even some war games the trainers conducted out beyond the solar system's planets. The three, fourteen-hour days of prep were exhausting but passed quickly.

On the Big Day, Erik couldn't believe the time was almost here. Launch was in an hour. Tightness gripped his stomach. He watched as the various crew members somberly arrived, each carrying a huge pack or massive foot locker containing personal belongings.

He noticed Marji talking with Jev. Her beau handed her a floral bouquet. The captain could almost feel her tears as she clung to Jev's neck for a long moment before they eased apart. Marji, her head down as she carried the flowers, dragged up the entrance ramp into the ship. Swallowing hard, Erik jammed his left hand into his pocket and touched the velvet ring box. He recalled his argument with Fred. He wondered: *"Why did I buy that ring? I'm anything but impulsive."* He had spent more money on the jewelry than a new Space Academy trainee earns in a year. *"I've been kidding myself. It's over and that I'll never see her again. Maybe this is what I've needed for closure."*

But another voice in his mind told him: *"Closure would be sticking that ring into a disintegration chamber. Or pitching it into the blackness of space. You know she could never love you. You're too different from one another.*

His thoughts flashed back to his youth on the dominant planet of Psi Serpantis, a multi-star system that included a yellow dwarf and several others stars. The main sun's companions were all within less than a tenth of a light year of one another. This made the system an ideal training ground for students seeking careers in the interstellar merchant marines, where a great-great uncle of his had worked, and in exploration. Cadets would make practice runs flying from one to another of the nearby stars. Erik's home world was the location of the large regional Space Academy that had been his alma mater. Several of his officers such as Montoya, Luci and even old Irv Malvo had all graduated from that school.

His attention returning to the present, the captain worked through a mental personnel checklist. When everyone was shipside, he turned on his heel and walked up the ramp to his new vessel, which he had christened *Freedom's Hammer*. The crew and he had barely had enough time to get used to the controls, defenses and weapons systems. But it gave the leader confidence to have the latest hardware at his fingertips. He made his way to the ship's bridge and plunked down in the main command chair, speaking into his comm link as he ran through last minute checks with Central Dispatch.

Montoya eased into a seat to his right. His eyes met the captain's. Fred's orbs looked like they could burn through Houston's. The captain steeled his eyes and stared at his second-in-command until the other man shifted his gaze.

Wiping away a tear, flight director Marji Faubner took her seat beside the other two officers. She initiated her protective ener harness.

Erik gave her a long moment. He reached over and touched her arm. "You okay?" he asked.

She nodded.

The captain received the go-ahead from Disp Central. "Okay, let's get this star blazer going!"

"Yes, captain," said Marji.

She eased the vessel away from the orbiting launch station then the flight crew engaged the engine.

12

Urgent Warning from Across the Stars

O N ANOTHER PLANET SEVERAL light years away, the science director was anxious to try out her new toy, the AV scope she had been seeking for years. It had remained low on the government's priority list not only due to the starship project but for other reasons as well. Leaders in the energy and telecom sectors had been clamoring for a space probe to monitor the activity of the larger of the binary system's suns. Severe solar wind and other bad space weather occasionally disrupted planetary communications and the local power grid. A few years earlier, the probe had been launched and sent into a solar orbit that kept it directly between the planet and its main star. The device gave ongoing reports that provided the planet with between fifteen and forty-five minutes' warning of any oncoming geomagnetic storms.

But the persistent official had finally gotten approval and funding to build the electronic eyes and ears that would allow the planet to gather a wide variety of audio and video signals from various interstellar neighbors and even distant worlds at the far end of civilization. Her technicians had tested the brand-new equipment and she wanted to be the first person to tune in live and eavesdrop on other planets. For this first run, she decided to try audio only. She brushed her hair back from her ears, donned a headset and began fiddling with the dials. Some bandwidths held only static caused by the background noise of cosmic radiation. A loud crackling made hurt hear ears and made her jump in her seat. Curse those sunspots!

She turned the dial. Ah, here was something. She strained her ears then increased the volume. A loud laugh track burst into her ear. Then a male voice and that of a female twittered on about something or other in an accent she found difficult to understand. A shiver went up her spine. Here was an actual entertainment show from somewhere out among the stars! But it would be nice if she could understand what was being said. Humans had been in space a long time and different languages and dialects had evolved. She plugged in the cosmic translator. Ah, that was better. Now she heard the speech translated into her people's tongue. She heard the words but the jokes didn't make sense. Maybe if she knew a little about the culture and could put the humor into some kind of context . . . She tried another program. Oops. Her face turned red and she felt flushed. Had to be an adult show. Some things were better left not translated. She flipped the dial a little further. Sports. This didn't work for her. She ran her hands through her hair. Her people had paid a small fortune for this?

The next station was playing some rapid instrumental music. She began tapping her foot, one shoe hanging off the end of her toes. This was good but she wanted some hard info. Random wasn't working. She entered some search criteria. The next sound was a professor with a rich voice lecturing about the life cycle of stars. Good stuff. She listened for some time, even plugging in her v-card to record it. She made a quick-note to flag the station for future reference.

Upon searching again, she caught part of a news broadcast in mid-sentence, laughing to herself that it wasn't really news. This signal had been flashing through space at the speed of light but the message had to be several years old, minimum. Her smile faded as the content of the broadcast registered in her mind. No, this couldn't be. She kept listening to make sure this really was news and not some sort of drama. No, it sounded real enough. She looked at the coordinates to note the location of the signal's origin. A nearby star system several light years away. She did another search. Had to be a different star system, the accent was dramatically different. Sure enough, the read out confirmed a location a few lights years away in a different direction. Another news program gave the same type of sobering report. She scanned another part of the sky. The same.

She was trembling. It had been over three years since the ship had left. Two years ago the vessel should have reached its first destination and sent a message announcing its arrival. The message would still be on its way back and would take another year before it could be picked up locally. The ship should have left for deep space almost one year ago. She enlarged the v-screen and pulled up an electronic log. She had previously agreed with the ship's captain to keep it on a schedule to make doubly certain she could get a message to the star travelers if needed.

She calculated where the ship should be on a certain date and positioned the scope's dish. She switched to broadcast mode: "May Day! May Day! Captain! This is Home Base. The ship is in imminent danger. Do not, repeat do not continue on to the red giant. Change course and return home immediately!" She knew the captain would expect details so she e-clipped the newscasts and sent those as well. She swallowed hard, realizing it would be a few more years before the ship got the word. She hoped she wasn't too late.

13

Counter-Attack

THE COLONY WAS LOCATED on the Starpath of Doom, the name the locals had given the flight path of the alien fleet. It was the only settlement that stood between the monsters and their destination, their home planet. The colonials had been dreading the arrival of the hostiles ever since they'd learned of the firestorm that had vaporized the Rantran colony. Area star bases had sent reinforcements ahead of the advancing fleet. The humans held war games then waited.

"Sir, the enemy is within a billion miles," reported a recon officer in a coded message.

"Cloaks on. Get the ships into position," ordered the commander in charge.

The two battleships and the destroyer engaged their cloaks. They moved into their assigned places and continued to keep watch. The aliens crept closer. Once they were within striking distance, the defenders' ships dispatched hundreds of nimble, cloaked fighter vessels. One fighter after another got an alien ship in its sights and opened fire. Space lit up with a number of nearly simultaneous explosions. A number of the invaders took hits. The fighters engaged in evasive maneuvers as the aliens shot back in the direction of the line of fire, trying to hit their unseen attackers. Despite their cloaking, several of the human fighters took hits, disabling some of the vessels. The aliens' ships hemorrhaged a number of fighters as well. One of the human battleships took a hit but its force field held. The humans circled back and fired again.

One of the human macro ships took one hit, then another but it managed to keep moving. The humans unloaded more punishment on the attackers. Several of their larger boats stopped moving. Despite the defenders' efforts, the massive fleet continued advancing. The still-cloaked humans continued firing at the flanks of the group as it progressed further into the solar system. The humans followed the ships and continued to pound them. A number of alien fighters broke off from the main group, created a vast spherical formation that surrounded many of the human ships. The attackers bathed the sphere's interior with fire, attempting to hit their unseen attackers. The action crippled several human fighters. One battleship took a major hit that disabled it. Its comrades offered assistance but the captain of the wounded vessel refused, urging them to continue their attacks on the fleet. The defenders had made valiant efforts and they had taken out a number of vessels but their overall impact on the invading navy consisted of mere pinpricks.

"Keep going!" ordered the commander of the human resistance. "We need to erode their capabilities." The protectors regrouped and leveled some additional damage.

When the invaders reached a half billion miles from the colony, two additional cloaked human battleships joined the fight. One of them soon took enough hits that it was crippled. The defenders eliminated more of the hostiles but the main navy forged on. The persistent humans kept up the fight.

"Casualty report," ordered the commander as the still-advancing fleet faded into the distance. The tally was two crippled battleships, fifty-eight disabled fighters, twenty-two destroyed fighters. Sixty-five known deaths, 205 personnel wounded. The officer put his hand to his forehead and sighed.

But because the humans had had the advantages of surprise and cloaking, the logistics personnel estimated the alien wreckage at ten times that of the defenders. But the armada was so massive the attack had barely slowed it down.

As their final offensive maneuver, the humans used the strategy the lunar sub-colony back in the Rantran system has executed, the surprise move that would not do much harm to the aliens for now but over time could weaken the invaders.

Hours later, their target insight, the alien fleet went into orbit around the inhabited planet. A half dozen local fighter vessels met them and engaged their weapons. The aliens ignored the pitiful display and unleashed a firestorm on the colony. Its deed having been completed, the enormous squadron moved out of the solar system. The human ships that were still functional mopped up the invaders in the wounded ships. But the aliens had gotten to re-sharpen their skills for the larger battle that lie ahead.

The starbases further down range continued to track the advancing navy. The enemy's flight path would not take it near any other inhabited systems. The brass would have been reluctant to stage an attack near a colony anyway lest the invaders wipe out yet another settlement. The humans continued to develop their plan, dispatching vessels to create another ambush. They set their trap in interstellar space, light years from the nearest star. The human leaders were again counting on the element of surprise plus expecting that by then the non-humans would have become complacent. Once the transient ships grew close enough the humans cloaked their vessels and opened fire. The attack quickly took out dozens of the boats. A protracted battle lasted for hours. The humans ultimately destroyed hundreds of enemy ships but in the end they barely slowed down the fleet, which continued to plow on through the reaches of space, seemingly unstoppable.

14

The Solar System of Death

THE SHIP HAD BARELY penetrated the outer reaches of the red giant's system when *it* happened. The vessel had passed through the
gravitational lens and had not even gotten as far as the Oort comet cloud
when the instruments picked up an abnormal space formation.

"Get me a visual," ordered Captain Houston.

Numerous objects, house-sized or larger covered a vast area of
space. The captain squinted. "I've never seen an asteroid belt so far from
a sun," he muttered. "Get me a closeup."

The view zoomed in on spheres of various sizes. "Something artificial," said the commander. "A defense system?"

"Some type of space mines," spat Montoya.

"Flight crew, steer clear of these objects at all costs," said the
captain.

"Yes, sir." Lt. Faubner responded. But moments later, she announced:
"Sir, some of them are following us."

"Deploy shields!" cried Houston.

Energy beams shot out from a few of the mines. They hit the ship
and rocked it despite the force field. The ship fired its laser cannons,
destroying several of the attackers. The vessel outran the rest but others
joined the pursuit.

"Cloak on!" said the commander.

Within seconds, the mines ceased attacking. Sighs of relief filled
the bridge.

Before long, sensors detected another type of foreign object. Closer inspection proved it to be the remains of a spacecraft, but one that did not appear to be of human origin. Over the next hour or so, instruments detected hundreds of ships of a variety of designs, apparent ruins from the mines' dirty work. The area was an interstellar graveyard. The captain ordered the crew to record all the ruins, a potential library of several alien races' ship building technology.

Erik shuddered. *"And to think we're barely into this solar system,"* he thought. He glanced around the room and suspected his fellow officers were all sharing that same thought.

"Have mercy," said Irv Malvo.

The other man nodded.

After some further travel the number of mines, and their victims, began to thin out.

Then Luci picked up a message another star system had broadcast years earlier to the general interstellar community. She immediately passed it on to the captain. When he heard it in his comm link, he hung his head. He audioed the entire crew. "Ladies and gentlemen, the aliens have struck again!" he cried. "We've just gotten word that they've wiped out another colony, one that was only a few light years from here, destroying a number of fighter ships and killing an estimated 4,800 people." An audible moan arose throughout the vessel. The captain went on: "This is a strong reminder of why our mission is so vitally important. We've got to succeed!"

Following the captain's message, Luci announced to the entire ship: "I hope that when the aliens have their war, the other side kicks their butts." Applause, cheers, hoots and hollers arose. But then the mood turned somber and no one spoke again for some time.

Finally, the instruments picked up another surprise.

"Great nebulae! There's a lifeboat out there!" cried Irv. "And sensors indicate there are survivors!"

"You mean there are aliens to capture?" snapped the captain.

Montoya, the ship's physician, shook his head. "The vital signs our scanners detect are quite human."

Houston wrinkled his brow. Headquarters had said human-inhabited worlds had told spacefarers to avoid this system at all costs. All media and shipping channels had been blanketing the airwaves. He ordered that the ship close in on the boat. Once *Freedom's Hammer* was close enough, a tractor beam pulled in the life raft. Within minutes, the tiny vessel was safely in the cargo bay. Houston and Montoya hurried down a hallway to meet the survivors.

As the rescued humans stepped through the cargo door and into the ship proper, the star men noted the height and deep tans character-istic of A'laamans. *"Have we fallen into a time warp?"* Houston thought. He had expected it to take several decades for the A'laamans to develop their own starflight capabilities, even with Daj's help.

"Welcome to our ship," the captain said.

The bedraggled survivors, five men and three women, thanked him profusely. Montoya did a med scan on each one. They were all in relatively good shape. Apparently, the boat's life support system was still functioning, a fact Marji and the other engineers soon confirmed. Houston ushered the refugees into the mess area. A crew man served them instant hot meals and beverages that the A'laamans wolfed down. Upon learning there were a total of six life vessels, the ship did an instru-ment search. They located all the others but none of the passengers on board were still alive.

"How long were you adrift?" Houston asked.

"Five, six hours," one of the ladies said.

"Seemed like forever," said one man. "Most of us had given up hope. We felt certain we were goners." Several of the others nodded.

"How many people were on the main ship?" asked Montoya.

"Sixty or seventy," a thin, dark-haired man volunteered between mouthfuls of grub. "Another 500 were with us originally but we left them on Paradise to start a colony."

Houston's jaw dropped. "Paradise" was the name the A'laamans used for that gem of an ELP that was in a neighboring solar system. He was irritated the planet had beaten the Association to planting a colony there. Once things settled down with this alien situation, perhaps some competition from a young, aggressive world like A'laama would get the brass back on their toes.

He continued: "Is there a chance anyone is still on the ship?"

A blond man with a brown beard said: "We urged our captain, First Officer Minj and his wife to leave the main ship but they refused to do it. They were afraid there wouldn't be enough boats for everyone. They stayed behind along with . . ."

Houston broke in. "Minj? Daj Minj? And Lisa?"

Several of the A'laamans nodded.

"We've got to find them!"

After searching the area for a time, the crew came upon one more battered ship, a vessel large enough to accommodate the number of people the A'laamans had described. "I have positive life scans of four people," the doctor said. "Let's go get 'em." The crew dispatched a pod carrying several members to round up the survivors. Montoya, and Luci, who doubled as a med tech, were among the rescuers.

As the pod approached the disabled vessel, the instruments indicated the living were all on the bridge and that the best route to them would be through the airlock from which the life boats had launched. The pod's computer tapped into the ship's and found the code needed to open the airlock. A circular door rolled back and the pod flew inside, docking in the empty bay as the door closed back up.

The star men donned force fields as they exited the pod. A second code opened the inner airlock door. Their hand-held instruments showed them a schematic of the ship. Wearing night vision goggles, they hurried down darkened hallways toward the bridge. After countless yards of corridors and several turns, they stepped into the command center.

The life support systems were barely functioning. The oxygen on the bridge was almost gone.

The temperature was 46 degrees Fahrenheit and dropping. The figures were slumped over in their chairs. Montoya swallowed hard upon seeing the condition of his old buddy, Minj, who was crumpled like a rag doll. Minj's arm was cold to the touch. Lisa lay beside him. Luci hurried to place oxygen masks on all four survivors while the doc took their vitals and administered some meds that could be absorbed directly through the skin.

This entire time, a 3-D kept playing showing a woman delivering an eerie message: *"The ship is in imminent danger. Do not, repeat do not continue on to the red giant. Change course and return home immediately!"*

Montoya, Luci and the others continued working on the four survivors. Once the doc was certain they were stabilized, the rescuers used some fold-up stretchers and anti-grav fields to move the A'laamans back to the pod.

Once the pod had returned to its mother ship, the rescuers transported their human cargo to sick bay. Montoya and Luci continued to monitor their condition. A few hours later, the evacuees were sitting in the ship's lounge enjoying a meal and beverages.

Minj was sitting in a hover chair verbalizing the trauma the group had just been through. "All those lives lost. And the ship. She was a beautiful ship."

The others nodded.

"That attack was so sudden," he continued. "We didn't know what hit us. The damage was done before we even deployed our shields."

"It happened so fast," the other male from the ship agreed. He was tall even by A'laaman standards and had thinning, white hair.

Lisa turned toward her husband. "Honey, you and the captain can't blame yourselves," she said.

The first officer lowered his head. She grasped his hand.

"You should listen to your wife," called a voice from the hallway. It was Captain Houston, who had caught the last part of the conversation. "That whole area is sown with some aggressive space mines. We barely escaped damage."

Minj turned around and stood in respect for his former commander.

Erik strode over to his former chief engineer. "Minj," growled the captain, hugging him and pounding his back. Erik next hugged Lisa, a short woman who had been dark-haired when he had first met her years earlier. A few silver streaks now highlighted her mane.

Minj said to Houston: "This is Mies Van Aaden, captain of our first ship, the *Independence*."

Houston grasped the larger man's hand and pumped his arm. "Glad to have you aboard," he said.

Erik stepped back and looked at his former chief engineer, who had been slim-waisted and buff while serving aboard the *Initiative*. He now had a slight paunch. His hair was thinning and flecks of grey highlighted his wispy mustache.

Houston looked up. For the first time, his eyes focused on an A'laaman woman who stood watching from the sidelines, the fourth person Fred and his assistants had rescued from the remains of the starship. Her beauty was striking despite the ordeal she had just been through. Erik did a double-take. She was tall and svelte, looking good in some form-fitting pants and a jersey top. She had light brown hair that fell halfway down her back. She looked so familiar but then . . . *it couldn't be.*

This *had* to be . . . *the daughter, Omma.* Or was this a younger version of the mom, a clone of her? The A'laamans had a custom of cloning many of their leaders.

He continued to stare at the lady, speechless. Her eyes met his then her gaze dropped to the floor. She quietly excused herself then stepped out of the room.

Houston stared at the floor for a long moment and balled his fists. His stomach felt on fire. After a minute he regained his composure and requested a crew member bring him a chilled glass of *calmjuice.*

Using a hand-held diagnostic, Montoya took another med scan of each evacuee still in the room. It looked like none of them would suffer any long term ill effects although the emotional trauma would probably linger for a while.

Once Montoya signaled Erik that the medical work was complete, the captain requested all of the A'laamans, including the ones from the life boat, meet with him in the mess hall. Once the weary group was seated, he began: "I'm sorry to learn of everything you've suffered . . . and for your losses. For those of you who don't know me, I'm Captain Erik Houston and our ship is called *Freedom's Hammer.* Years ago, I visited your home planet. I love the A'laaman people and your culture. I want you to consider this ship your temporary home. It will be a little crowded. But we'd be glad to share our meager accommodations and food with you. I've instructed my crew members to treat each of you like one of us. But it's important for you to know that we're not yet out of danger. Before we can return to our home base, we must drive deeper into this solar system and conduct a recon mission."

A'laaman Captain Van Aaden snapped: "I've just lost fifty-four of my people to these barbarians! Who knows what other perils lay ahead? And you're telling me we need to push on?"

Houston waited a moment before responding. "Captain, I can appreciate how you and your people must feel but we're not here to conduct a rescue operation. My first priority is to follow our explicit orders. The aliens who inhabitant this system not only set the traps that destroyed your ship and killed many of your people. They also wiped out an entire colony of humans less than ten light years from here and another only a few light years away. We have a limited window to conduct our mission because this area will shortly be the site of a war between two competing alien races. I hate to see you dragged into this situation but if we can survive the next day or so, we'll be able to get all of you to safety."

The other man glared at Erik but held his peace. Houston glanced around the room. The wide eyes and slack jaws made him feel he should have waited until morning to break the news. But by then, the mission would be underway.

Daj stood up and broke the tension. "This isn't just *your* battle. This is *humanity's* battle. I'll go planetside with you," he said.

Lisa stared at her husband with dagger eyes.

He put his arm around her shoulder. "Now, honey this is an emergency," he said. "This is something I just have to do."

A few of the A'laamans, including the mystery woman also stood up and offered their help.

Houston held up his hands. "I deeply appreciate your support, but all of you have been through enough already. And Daj, I can't come between you and Lisa."

She grasped Minj's hand. "I support my husband's decision," she said. "He's right. We need to do whatever we can to stop these aliens. Just . . . bring him back safe, Erik."

"All right," said Houston. "Thanks. Both of you. I'll gladly accept Daj's assistance. He's the only one of you who's received comparable training to my crew and I. We need to discuss coordination of efforts. We begin operations tomorrow morning at 0500 hours."

"Before that," said the A'laaman captain, "I need to ask your indulgence. We've talked among ourselves and have come to a consensus. Given all the trauma we've just been through, it would give us a sense of closure if we could have a memorial service for our people."

Houston felt a huge lump in his throat. "Yes, yes. Of course. Let us know what we can do to help."

Sometime later, the guests held a simple service for their fallen comrades. Erik and as many of his crewmen who were not on duty also attended. The beautiful A'laaman woman stood in the crowd across the room from Erik. He tried to make eye contact but she did not look his direction. Van Aaden briefly spoke, naming some of the more prominent of those who had perished. The group said a prayer then sang a hymn. The survivors spent some time hugging and consoling one another.

Shortly after the memorial, security chief Irv Malvo was on duty when The Woman approached him. "I-I need to see Captain Houston." she said.

"Ma'am, we're glad to have you A'laamans on board but certain areas of the ship must stay off limits to you. This is the entrance to the bridge, the command center. Not even our crew members are allowed here except when they have official duties."

She nodded. "Then could you please get a message to the captain?"

"What would you like me to tell him?"

She was silent for a moment. "It's personal. Tell him I need to talk with him."

Irv rolled his eyes. "Ma'am, we're about to undertake a life and death mission."

She laid her hand on his arm and looked at him with earnest brown eyes. "Please, sir. I think he'll really want to hear what I have to say."

He sighed. "Does he know you? Who shall I say wants to meet with him?"

"A friend."

He wrinkled his brow. Waving his hand, he said: "Alright, missy. You seem sincere and this sounds important so I just sent him a message. He's got pressing business to attend to now but I'm sure he'll get to you as soon as he can."

"Thank you, sir." She turned on her heel and walked away.

Two hours later, Erik was striding toward the ship's lounge. Once Irv had e-linked him a 3-D of the woman requesting the meeting, he knew he would never rest until he got this cleared up. He saw her at a table and plopped down at a seat across from her. He waved his hand to set up a sound and visual block. He looked her. "You wanted to see me," he stated, looking into those orbs.

She touched his hand and returned the gaze. "Captain . . . Erik . . ."

His gaze intensified. Silence reigned. Finally she said in a low voice: "I'm Zama."

Her words knocked the breath out of him.

"Look. It really is me." She rolled up her left shirt sleeve to reveal a pink, starburst mark on her shoulder. She pulled on a dainty gold neck chain, pulling the rest of the chain out of her shirt. The chain ended in a small, gold cross.

The captain's jaw dropped. She had apparently had surgery to remove the bump on her nose. Her figure was considerably slimmer and she'd lost her double chin. She had been forty-four Standard years old when he'd last seen her. It had obviously taken Daj some time to help the A'laamans develop the starship. So how old must she be now? Early fifties? Mid-fifties? Older? He shook his head. She barely looked thirty-five. Her skin was flawless other than some laugh lines around her eyes.

Seeming to read his mind, she said: "Now captain, does age really matter between friends?"

He giggled like a lovestruck teen.

"A lot happened after you left A'laama," she continued. "At the time you knew me I was under fire politically. There were the assassination attempts . . ."

He wrinkled his brow and nodded as he recalled helping her through those difficult times.

She continued: "After we broke up the plot against me, my popularity went back up. I got re-elected in a landslide. I felt really blessed. I got back to running and working out . . ."

He glanced at her left arm and noticed several tiny red and blue lights flashing under her skin due to the bio implant that regulated her heart meds. Her heart trouble pre-dated his visiting her planet. "You look incredible."

She was beaming. "Thanks. You look great, too. Of course, you haven't aged since I last saw you. Well look, I just wanted to say hi and let you know I'm here." She stood up. "It's really good seeing you again."

He stood and held out his arms. She hugged him but they barely touched. She turned and walked off, leaving the captain shaking his head.

A short time later he strolled down the hall to the room where Luci bunked. After he knocked, she stuck her head out the door. She was wearing lounging pajamas.

"Is Zama here?" he stage-whispered.

The crew member nodded. "I set up a separate hover field for her and showed her how to adjust the comfort setting."

The door closed and re-opened a minute later. Zama's hair was in bangs. Dark circles underscored her eyes. She smiled when she saw him. "Hi," she said.

"I brought you something." He pulled her old overnight bag from behind his back.

Her thoughts flashed back to what she had been feeling years earlier when she had packed the bag. Her face turned red. "How did *you* get this?" she cried. "My personal things are in there!" She grabbed the bag out his hands and pushed the door shut.

He swallowed hard and shuffled away.

15

Starry Night

Z AMA DROPPED THE BAG in a corner of the cabin and plopped into a seating position on the hover bed.

"Uh . . . are you okay?" asked Luci, who was lying on her own hover bed, a two page spread of e-magazine text and 3-D photos hanging in mid-air a few feet away from her.

Her guest nodded. She sat there silent for some time then grabbed the bag, strode into the cabin's restroom and closed the door. She opened the bag and began sifting through the items. Given that her fellow A'laamans and she had escaped with their lives she suddenly felt wealthy as she poured over the amenities. She held up one of the bottles of sparkling beverage. She could use a drink. Maybe later.

She pulled out some of the cosmetic items, thinking that if she were able to feel clean and pretty again it would lift her spirits. She picked up a cylinder the length and thickness of one of her fingers. She held the instrument close to her face and turned a tiny dial. Countless nanobots, too tiny to be seen, flowed onto her face and massaged it, removed dead skin cells and cleansed her pores. Within a few minutes her skin was pink and she felt as if she were glowing. She turned another mini-dial. An army of nanos poured across her lips and into her mouth. Each round micro bot quickly revolved, scrubbing her teeth and massaging her gums. By the end of the session, her mouth had a fresh, sweet taste. She turned a third dial and moved her finger from left to right in front of her lips. A new tiny brigade used microscopic brushes to paint on her favorite shade of lipstick. A human observer would see the color appearing on her lips with the wave of her finger as if by magic. She gazed into the mirror and squinted at her lips. She moved her finger in the opposite direction and the color went away. She moved the dial to another setting and repeated her original motion. A different color spread across her lips. This one met her approval.

She closed the bag and went back to the sleeping area. Her host was curled up, apparently out for the night. Zama laid back on the hover bunk and tried to make sense of the day's events. What was she doing here in this strange solar system, a shipwreck survivor? Why had the starman come back into her life? She knew God had led and guided her throughout all her years. What kind of crazy plan did he have for her now?

"In the beginning, man lived on a single planet. A tiny, blue orb known as Earth. And over the millennia, man developed technology that freed him from the bonds of gravity. Generation after generation, humans reached farther and farther into space, spreading their seed among the stars. One colony begat other colonies, and those colonies begat other colonies still. And man began to multiply among the stars . . ."

Zama drank in of the deep, clear voice that spoke these words. She pictured in vivid detail everything the voice described. An expansive, blue world swaddled in white sat before her. It gradually shrank to a dot before vanishing in the distance. In all directions, many stars broke the blackness of space. Big, bright stars, Tiny, distant stars. More stars than she had ever seen. More stars than she could count. She stopped breathing for a moment. She drew closer to one of the stars. Then she began rushing toward it. She zipped past the star and drew near to another. It expanded until it took over her entire field of vision, until everything was a blazing ball of nuclear energy so bright she shielded her eyes with her arm. She brushed past that star and hurled toward another. And another. And another. Many stars began rushing by. Faster, faster. The stars became a blur, melding into rushing streaks of light. She felt as if *she* had become light. *She* had become energy.

She awoke with a start. She was breathing heavily and her heart was pounding. She wiped her damp brow with the back of her hand. Glancing around the darkened room, she remembered she was still on Erik's ship, still lying on the hover field and trying to sleep. What had caused her to re-experience that vivid dream, the one she'd had many years ago as a teenager who had fallen asleep while lying on the roof of her parent's home, gazing at the stars? Her thoughts wandered back to that long-ago time. When she had awakened, the wind had tousled her hair. Despite the coolness of that night she'd felt warm all over. The teen had gazed up at the stars, so few and dim compared with the ones she'd seen so vividly just moments earlier. She knew they were some of the same stars but she had seen them from a different time, a different place.

"What does this all mean?" she'd whispered.

After the first time she had the dream, she had a strong feeling that her people needed to go to the stars. She shook her head, wondering how this could be. Centuries earlier, people from the stars had colonized her planet. But their ship had crash landed. Most of the would-be colonists had perished in the crash and much technology had been lost. Her people did not know how to star travel. They did not know how to go into space at all. Yet she kept feeling that was her people's destiny.

The teen had slowly nodded. "The God who made the stars can get us to the stars," she'd said, barely audible.

She'd had a difficult time going back to sleep that night. The vision had haunted her thoughts off and on for the next several days. But eventually she got busy again with her public speaking classes, trying out for the school play and spending time with her boyfriend. As the years passed, she forgot the vision. Many years later, well into adulthood, she awoke in the middle of the night. She had a vision like the one she'd had in her youth. She rolled out of bed clad in her nightgown and shuffled onto the balcony, her bare toes curling in response to the cold concrete. She looked at the stars, brighter than normal. She realized her people *must* go to the stars.

After that night, she had a nagging sense she needed to do *something* to help accomplish the mission, but it was never clear what. At times she would pray to ask what she should do. Her other prayers seemed to get answered. But ones about the stars seemed to lack a response. Finally, she forgot about the dream again.

By that time, she was an attorney. As her career progressed she became a judge, a legislae, then finally chiefexec of the planet. Over all those years, the vision did not recur.

One evening, Chiefexec Zama Elle was sitting at a desk in her private study, reading some on line reports. Her personal v-screen, the size and shape of a small, hand-held makeup mirror, abruptly turned crimson and began flashing. Her stomach tightened and her mouth was instantly dry. A Red Alert! The first one ever during her administration.

"Yes," said the chiefexec through gritted teeth.

The screen stopped flashing and returned to normal. The face of her science director, Tanna Bern filled the screen. "Madame Exec! An alien space craft has entered our solar system!"

A rush of thoughts flowed through Zama's mind. Her culture used the term alien in a broad sense so she knew Tanna likely meant a ship full of humans. *Anyone* who was not of their planet was an alien. The chiefexec remembered her twice-seen vision from years ago. And the confusion over what, if anything to do about it. And this approaching ship . . . was this the old Angel's Prophecy coming true . . . people from the stars arriving just short of 300 years after the colony's founding? The planet had been isolated, like an island is space. Her face became flushed. Her hands felt clammy. She could feel her heart pounding in her chest.

The adrenaline-pumped Zama slammed her hand on her desk. "No space aliens are landing on my planet!" she snapped.

"We may not have a choice," said Tanna. "Do you want to anger some interstellar authority that has starflight capabilities? They could also have some powerful weapons."

The chiefexec bit her lip. She glared back at the 3-D image of her planet's chief scientist. Tanna was thin and had straight, medium brown hair that abruptly ended just past the nape of her neck. "You're sure it's a ship?" the leader replied.

Tanna looked pale. "Yes. The past several days, our astrogazers have been tracking a small object. It started out in deep space and is moving toward us at a significant fraction of the speed of light. It is gradually slowing down and should arrive here within two days."

Now the thump-thump of her heart felt even more pronounced. She could feel her blood pressure rising. She took a deep breath, certain her arm implant was already shooting meds into her bloodstream. "And . . . there's no possibility this is some *natural* phenomenon?"

"Not something moving that fast, Ma'am."

The leader sighed. "Looks like we'll be getting some visitors," she finally said, her voice trembling as she tried to look nonchalant.

She later learned that an interstellar association had sent the ship. They considered the planet to be a lost colony. The ship carrying the ancestors of the planet's inhabitants had crashed light years short of its destination so there was no official record that A'laama ever had been colonized.

The guests stayed for several months and shared considerable tech knowledge with the planet's natives. But the captain, Erik Houston had laughed at the chiefexec when she had asked for help in building a starship. He implied the planet's science was so primitive he would hardly know where to begin.

Yet, the ship's chief engineer had fallen in love with one of the local women and had stayed behind to marry her. Would Zama's vision of the stars come true after all?

Her thoughts jumped forward several years to a time when her planet's starship was still in the development stage. She had been on a dinner date aboard a zeppelin floating thousands feet above the city. The metropolis had been aglow with millions of tiny lights. The elegant restaurant known as the Zep had only been open a few months and was already the rage across A'laama.

Zama, clad in a sequined evening gown and lacy shoulder wrap, gazed out the window at the sight. She was basking in the glow of that genius Minj. He'd finally done it. Discovered some scientific maxim that would cut years, maybe decades off production of the starship. The media had been so elated they'd even named his discovery the Minj Principle. Daj had thought they could have a working prototype within a year.

"What's wrong? You seem far away tonight," said a resonant male voice. He reached out and clasped her hand.

She came back to her surroundings with a start. "I-I . . ."

She looked into the eyes of her beau, John Clayton, a dashing figure with thick, white hair. He was a senior *legislae* and longtime political ally. He had been seeing Zama ever since his wife had divorced him over a year earlier.

"You've hardly touched your meal. Is everything all right?"

She nodded but didn't look convincing. The youthful ex-chiefexec had brown hair halfway down her back. She was wearing gold earrings and her favorite piece of jewelry, a dainty, gold cross necklace. The scent of her herbal shampoo wafted through the air. "It's the ship," she said. "I'm thrilled that it's finally coming together."

"Yes, the ship," said Clayton. "Honey, I . . ."

"Yes, John."

"You've worked hard on that project. I'm thrilled with the progress you've made. But I've got something on my mind, too. I've decided to run for chiefexec in next year's election."

She began breathing deeper. "*You* decided. But . . . I thought *we* decide things."

He continued: "I didn't want to bother you with this at first because you've been so focused on the ship. But I can win this election. I could really use your help with my campaign. Then you and I can continue working on the planetary exploration programs you initiated."

She turned back toward the window. "John, I've got something I need to tell you, too." She faced him once again. "I've shared with you the vision I had years ago. The one that's compelled me to pursue the ship."

"Of course."

"Well, the last few days I've had a strong feeling that won't go away. I-I need to go on the maiden voyage."

"You *what*?"

"I need to be on that ship."

"Why, because *God* told you?"

Her stomach had knotted at the way he'd pronounced the word "God". "No, he hasn't *told* me. But I believe he's *leading me* to go to the stars. I've prayed about it a number of times. I think it's what he wants me to do."

Clayton's face turned red. "That's the craziest thing I've ever heard."

Zama looked down. She wondered how she could think she was in love with this man when he seemed to place so little value on her faith. She looked him in the eye and said firmly: "This is something I *need to do.*"

Clayton's eyes became burning flames. "Y-you didn't used to be like this. Years ago, you were content to just live in the Crystal Mansion and run A'laama. Now, you've got to go explore the *outback* and take trips into *space . . .*"

"Maybe I've grown a little."

"It's Houston, isn't it? You were never like this until he came here. You've still in love with that Erik Houston."

She whipped her head around, sending her hair flying. "How dare you?" she cried. "I quit caring for him a long time ago. I'll die of old age before I ever see him again."

"Well, he put some crazy bug in you and now you're never satisfied. You always want to do more . . ."

She swallowed hard, knowing there was some truth to his words. Something had begun to change inside her once she had gone with that spaceman into the wilderness.

Clayton tried to put his arm around her but she shook her shoulders and he retreated. She turned back to the window and froze him out the rest of the evening.

Once the Zep touched down, she arranged for her own transport. Instead of going home, she showed up the house of her chief advisor and best friend. A half-comatose Kelly Farji answered the door, clad in a fuzzy robe and fluffy slippers. The full-figured woman's long, dark hair was in disarray.

"Do you know how late . . ." Kelly spied her friend's tear-stained face and swung open the door.

Zama threw her handbag onto the plush couch then bounced over beside it. As the two women shared a fragrant pot of javamon, Zama launched into a monologue. "I know what you're going to tell me," she said. "You're going to say you knew from the beginning it wouldn't work out with John."

Her counterpart looked at her and opened her mouth, but Zama continued: "But why wouldn't it work out with him? We're political clones, we've always gotten along great, he's a wonderful dancer, he's handsome, he's one of most desirable men in town. But . . ." She stood up and began to gesture as she paced the floor. "He's not the One. He's not the one named in the Prophecy. Well, curse the Prophecy. I don't want to be like Wendi in that ancient fable, Pitr Pann, an old lady waiting on a man who's hardly aged. I can find my own boyfriends. Curse the Prophecy!"

Kelly looked at her visitor. "Sweetie . . . I know you're upset . . ."

Zama sat back down and looked at her friend. "I-I'm sorry, Kel." She buried her head in her hands and began to weep. The host sat beside her guest and placed her arm on her shoulder.

"Why do I have such bad luck with men?" Zama sobbed. "I loved Andreu but he died twenty years into our marriage. Erik and I got close but then he had to go back into space. Now I find out John hates my devotion to God and puts his career ahead of me."

Kelly handed her friend some additional tissues and continued to listen. And Zama knew that she would need to go on despite her broken heart. The stars beckoned.

As her memories of that time kept flooding back, the former ruler of A'laama rolled over on her side. A few minutes later she switched to another position. She made a conscious effort to go back to sleep. After some time she was still unable to relax. She glanced over at Luci, several feet away on her own hover field. Her host's form seemed motionless in the darkened room. Zama eased to her feet and with quiet, cat-like motions slipped out of the cabin. The anxious woman began to wander the hallways of the ship. She set up a sound block and pulled her v-card from her pocket. Pushing a tab, she summoned Virtual Kelly. A 3-D of her friend from A'laama appeared. Zama had had the real Kelly's image, personality and memory uploaded to serve as a traveling companion on the long trip. She poured out her grief about all of her people who had perished in the attack on the A'laaman ship. She had known some of her cohorts for many years. She detailed her heartbreak over the ill-fated ship. Then her thoughts turned to the shock of Erik Houston being thrown back into her life.

"I don't know what to do," Zama began. "I don't how I feel after all these years."

"Take your time," said her virtual friend. "Get to know one another all over again."

"But . . . these crazy circumstances. I'm a ball of stress. Even if I still had feelings, how would I know they're real?"

"You both had a lot of stress on A'laama. Were those feelings real?"

"*That* was light years ago. In more ways than one."

Zama swallowed. She continued to talk with Virtual Kelly for a while then, with a snap of her fingers, V-Kel retreated to the *holoworld*.

The lady continued to walk the halls. Finally, she came to the bridge. She looked both ways. This time, there was no security detail. The door was wide open. All overhead lights were out. Erik sat in the dark, occupying the great chair at the command center of the powerful ship. He was gazing out the port at the sea of stars, one hand cupping his chin. She crept through the doorway and gradually eased closer until he she was standing just behind him.

He said quietly: "You realize, madam, that I could have you court martialed."

The shock of hearing his deep voice break the quiet made the back of her neck tingle. After a moment to recover, she replied in a low voice that unnerved her when she heard it because it sounded unintentionally sultry: "No you can't. I'm not in that Navy of yours."

Silence.

She hoped he hadn't taken her response as a sign of disrespect. She cleared her throat before speaking again. "Listen, I-I'm sorry I snapped at you earlier."

"That's okay," he replied. "You've been through a lot."

"But there's still no excuse for the way I acted. That was sweet of you to bring my overnight bag to me but . . . I haven't seen it since you were on A'laama all those years ago so how . . ."

He nodded. "You left it on our ATV, which ended up back on my old ship. I found it and have kept it with me because it reminds me of you. I was hoping some day I'd see you again. But I never thought it would be *here*. Or under these circumstances."

She swallowed hard. "Life has a way of playing jokes on you," she said. "I'm the only A'laaman on board with a change of clothes but it doesn't do me any good because I was heavier when I packed that overnight bag years ago. Everything I try on falls right off me."

He shook his head and chuckled. When the laughter faded Erik, Zama, and the stars faced an awkward silence. "I can't pretend to be a hero," he finally said. "When I found your bag, I looked through your stuff."

"You what? Erik!"

"I know they're your personal things and it was none of my business. I'm sorry."

She sat down on the floor beside his chair. "That's not the worst of it. Now you know what was going through my head when we went into the wilderness. I feel like such an idiot. I was lonely and I was attracted to you and . . ."

He shook his head. "Don't feel bad. I was struggling with the same feelings for you. But . . . we didn't do anything."

"No. No, we didn't." Silence reigned for a moment. "A lot has happened since then," she said, gazing out at the stars without really seeing them. "You weren't there to see my daughter, Omma start her own cosmetics company. You weren't there to see my son, Andy become a major force within our starship program. Whenever we build a second ship

. . ." her voice broke, "he hopes to become the captain. You weren't there for the marriages of my kids or the birth of my grandchildren."

He swallowed. "G-grandhildren?"

She nodded, wondering if she would ever see them again.

He placed a hand on her shoulder.

More silence. Then she changed the subject: "Thank you so much for rescuing my people and taking us in. Your kindness means a lot to me."

He lowered his head. She couldn't see him blushing in the darkness. "You would do the same for us," he said.

She held out her hand and he helped her to her feet. "When you land on that planet tomorrow, you make that mission happen. We're all counting on you."

"I won't let you down."

She patted him on the shoulder then turned to leave.

"Zama," he called after her. She turned around. "I'm glad you're here."

She smiled and walked off into the night.

16

Light Years from Paradise

Daj Minj was another shipwreck survivor who was finding it difficult to sleep that night. Although he was not a native of A'laama, he grieved the loss of life and the destruction of the ship. He had come to the planet years earlier and had stayed behind when Captain Houston's former ship, the *Initiative,* had completed its mission. As the ship's chief engineer, Minj had brought to the technologically backward world the expertise on how to build a stargoing vessel, something planetary ruler Zama Elle had strongly desired. The starman had carried with him an info chip holding a comprehensive history of starflight techniques over the centuries, from the old generation ships to modern near light speed craft. The first starships had been tiny, unmanned nanoships. Eventually humans had begun exploring the stars. The chip covered all star travel technologies: ion drives, nuclear power, solar sails, ram jets, space warps, antimatter engines. The chiefexec had spent the political capital, drummed up the media support and arranged much of the financing to get the job done. He had supplied much of the brain power and perspiration.

Also helping make the starship a reality was Daj's passion for long defunct sports of ancient cultures. He had loved hoopball, known in a bygone era as basketball and had introduced it to A'laama. He had considered the game a natural fit for such a small planet that had gravitational pull of just .78g. The natives tended to grow taller and could jump higher than on many other worlds. A fast-moving game of hoopball offered the chance to see star players dunk balls through hoops that were a full fourteen feet off the ground. Daj had organized separate men's and women's leagues that had proven wildly successful. As founder and part of owner of all the franchises, he could have amassed tremendous personal wealth but instead had donated all the profits to help fund the space effort, keeping the Starship Tax within politically acceptable levels.

He had become increasingly excited as the ship began to take shape and the launch date approached. An amateur historian, he had noted that A'laama orbited the star Eta Cephei, an orange giant, and its dwarf companion star. The Arabs on ancient Earth had called the larger star Al Agemim or Al Agimim, meaning "the sheepfold". Minj speculated that the colony's founders had been aware of this but over time the old Arab term was corrupted to refer to the planet itself and became "A'laama", which in local usage came to mean "snuggle or nestle". Part of the orange star's official name, Eta evolved into Ava while the dwarf star, A'laama's other sun, came to be called Zoe.

The starship's first journey would be to a planet orbiting 18 Scorpii, a yellow orange star. Astronomers on ancient Earth had located the star at the northern edge of the constellation Scorpius, just off its left claw and forty-five and seven-tenths light years from Earth. But for the A'laamans, the star was a near interstellar neighbor. The planet had been the intended destination of the A'laamans' ancestors but their ship had malfunctioned short of its goal, forcing it to crash land in the Eta Cephei system on the planet that later became known as A'laama. So the modern day descendants now wanted to "correct" history by planting a colony more than three centuries after the original attempt. 18 Scorpii was considered a twin of Earth's Sun, having about 1.02 times its mass and 1.05 times its luminosity. The destination world was a gem of an orb, the most Earth-like planet (ELP) that had ever been found. The A'laamans had been unaware of all this background until Minj had shared it with them.

The A'laamans had always referred to the 18 Scorpii planet as Paradise. The starship crash survivors had been aware of the lush status of the world that fate had denied them. Adding to their early despair and frustration were the living conditions on their new home planet. A'laama was intensely hot and bathed in heavy UV radiation that gave most of inhabitants a deep tan and a higher than normal risk for skin cancer. Four years in, the colony almost died out due to a lengthy drought and the resultant crop failure and famine. At that time, an angel appeared to several colonists and spoke the Prophecy to give them hope and confirm that the drought would soon end. The colony would survive and eventually prosper.

But by then the idea of the other world in the neighboring system being a paradise was even more ingrained into the A'laaman psyche. Musicians had written folk songs about how the faulty ship had condemned the colonists to a less than ideal existence. One of the more popular tunes was called *Light Years From Paradise*. Over the centuries, it had evolved into a drinking song rendered with great emotion at pubs all over A'laama. Daj was a private man. Other than developing the ship and attending some of his teams' hoopball games, he spent most of his time quietly with Lisa. But he was fond of the mild intoxicant known as Amba Stout and would occasionally down a frosted mega glass or two at a local pub and could sing a rousing chorus of *Light Years From Paradise* as if he were a native.

Daj and the A'laamans had designed the ship for both colonizing and exploration. The vessel, the *Independence* carried its inhabitants to the Paradise planet where about 500 of them remained behind to found a colony. The new colonists used high-powered lasers to carve a memorial deep into solid rock. Captain Mies and Zama had declared that event the official start of the Interstellar Republic of A'laama, beaming a celebratory 3-D message back to their home planet. The republic would for now consist of the two planets, A'laama and Paradise.

Scientists back at home base had insisted that the ship be used more than once so several dozen people had remained on board for its second voyage, the trip to do some closeup studies on the red giant, the star that just happened to be the sun for the home planet of the fierce alien race that had destroyed the two human colonies.

17

Cosmic Math

MONTOYA WAS SHAKING HIS head. The leader of that crazy planet showing up on a failing starship! How could that be? The first officer couldn't quite grasp it. He moved his fingers through the air. Given the estimated population of human civilization, what was the likelihood of two individual star travelers several light years apart running into one another at a specific time in a specific place within a specific star system? A light blue florescent decimal materialized in answer to his question. There was a long string of zeroes to the right of the decimal point, followed by several digits of numerals. He stared for a long moment then repeated the calculation. Same number. He changed a couple of the variables and got an even smaller number. The first officer frowned.

He looked up the estimated number of starships then the number of solar systems in the tiny corner of the galaxy humans had explored. What was the chance of a disabled ship with a limited window left on its life support being discovered in time to save some lives? Especially when the discovery was made by a vessel from another star system that just happened to be passing within a stone's throw of the wounded ship? Another nano-number. Infintesimal.

How could this be? This made about as much sense as Daj's personal history. Eons ago, he had been married for a few years to a lady named Nanci while they both attended the Space Academy. The couple had divorced then briefly reconciled. During their last days together, Nanci became pregnant. But the couple then split up one final time and went their separate ways – Nanci to a colonizing ship and Daj to a First

Contact vessel that followed up on previously colonized worlds. Nanci was on the ship that crash-landed on A'laama. She survived the crash. Three centuries passed on the planet.

Meanwhile Daj, spending much of his career aboard ships traveling at near light speed had only aged about fifteen years, thanks to the time dilation of Einsteinian Relativity. Finally, Daj's ship, the *Initiative,* was dispatched for a First Contact to A'laama, where he learned that 300 years earlier, Nanci had given birth to their daughter, who had become an ancestor to many modern-day A'laamans.

Recalling these things, Montoya pulled on his beard. He had personally developed the DNA results proving the baby was Daj's child. Now what was that probability? The starship officer did the math. All of these things, virtually impossible.

Until now, he had believed in the old folklore about Einstein's Law, an updated version of the ancient Murphy's Law: that Einsteinian time dilation caused some weird and unpredictable effects, some remotely probable events that could not be explained any other way in a universe that otherwise seemed quite random. But this latest, this Zama showing up out of the blackness of space . . . Was it just that she annoyed him, was that what bothered him so much?

He gazed out the port at billions of jeweled stars and the vast sweep of the Milky Way. Maybe has belief system was all wrong. *Was there a God after all?*

18

Preparing to Engage

THE SHIP CONTINUED TO progress deeper into the hostile solar system. It passed through the comet cloud. Later, it flew by a dwarf planet, a frozen ice ball. Eventually, another. Next, it approached a gas giant.

A silent alarm woke Erik in the middle of the night. He stumbled to the bridge. Montoya and Irv followed quickly behind him. The instruments had detected another menace-- a substantial navy patrolling this portion of the solar system. He studied readouts on the ships. They appeared to all be drones. One of them was only 30,000 miles off. But *Freedom's Hammer* had continued to deploy its cloaking and its shields. He ordered his two officers back to bed then headed to his own cabin.

A couple hours later a series of rapid pounding and banging noises jolted Erik awake. Rushing to the bridge, the captain saw what appeared to be countless objects slamming into the port, ruining visibility. "What's happening?" he cried. "Have we lost our cloak?"

Irv did a scan. There were, indeed, thousands of small objects pelting the ship and were it not for the shield the missiles would have reduced the starjammer to space junk. The noise sounded like a cloudburst pounding on a tin roof like the ancients had used milenia ago.

"What is this, some kind of cosmic hail?" snarled Montoya. "The Association should have sent an entire fleet for this mission."

The captain shook his head.

"You're just one small vessel trying to fly under their radar," said Minj.

"Fly under what?" asked Montoya.

"In ancient times . . ."

"Not now!" snapped Houston.

As the officers further studied the scan, they realized the ship was not being targeted. The rapidly-moving objects filled space for thousands of miles. It was a hailstorm belt, apparently yet another line of defense. The ship remained unscathed until some bursts of energy struck. A strong explosion slammed into the shields and rocked the vessel. The lights went out, causing anxious gasps and cries from all areas.

"Engineering!" Houston shouted in the dark.

"On it, sir," snapped chief engineer Marji Faubner.

Moments later, the lights came back on. Ener waves pounded the ship for several tense moments then subsided.

"Damage assessment," called the captain.

"Sir, there's has been damage to some of the e-systems, particularly the lift shaft," Marji replied.

"Get on it! Show time's just a few hours away."

"Yes, sir!"

The vessel sped deeper into enemy territory. Marji continued to work on repairing the damage. She ordered the assistant flight director to guide the vessel into orbit around the ringed planet that was the stamens' destination. Other crew members arrived at their posts while most of the traumatized A'laamans tried to sleep.

The comm link in Erik's ear indicated the cloaking device was still engaged. That should allow them to operate without alerting the natives for at least a while. A message from Luci indicated the ship was scanning for communications and had found some chatter coming in toward the planet and other messages going back out. The mathematical de-scramblers were working on the messages. Yes. A battle was coming soon. Erik and his crew had barely gotten there ahead of the two alien navies. Houston clenched both his fists until his fingernails dug into his palms.

A scan of the space surrounding the planet revealed dozens of large, heavily-armed space stations, each housing a half dozen small ships, probably fighters. A schematic displayed blinking lights of various colors showing an intricate communications web among the stations and between them and numerous points on the planet's surface. The captain whistled. If the enemy ever detected the cloaked starman vessel, it would all be over within minutes.

"We're really in the lion's den now," said Minj, coming up with another historic/cultural reference no one but he understood.

"Uh . . . Captain, this is Lt. Faubner," said the comm link in his ear.

"Yes. Is the lift shaft up and running?"

"Just about got it, sir. A few more minutes."

"Good."

"Sir, we've found something else of concern."

"Go on." The captain listened, his forehead wrinkling in concentration. When his chief engineer finished her report, Erik lowered his head and placed his hand on his forehead. From the beginning he had labeled this a suicide mission. He thought: *"Lord, my life is expendable. But I hate to lose my crew. And what about the A'laamans? They're like innocent lambs who wandered into this."*

Sensing the captain's mood, his second in command asked what was going on. He turned toward Montoya and told him.

The first officer grimaced. He arose from his seat and began pacing the floor. "Erik, we're going to need a miracle to come out of this alive," he said in a low voice.

"Fred, if there's ever been a time when I've needed you to have a little faith it's now," he replied. But silently the captain was praying, *"God, please help us get through this or we're all going to die!"*

He placed comm calls to several of his crew. "Virtual officers' meeting, immediately," he snapped.

Within moments, head and shoulder 3-D's of Irv, Luci and Marji all appeared. "Lt. Faubner reports something of grave concern," the captain began. "The entire planet is surrounded with an electronic field that serves as a security system. Any object beyond a certain mass, even as large as a vehicle or a human body, that penetrates the planet's atmosphere will trigger an alarm that will go off across the entire planet. Ms. Faubner, is there a way we can execute our plan without triggering the alarm?"

She stuck the end of her ponytail in her mouth and chewed on it a moment. Removing the hair from her mouth, she said: "I can work with Luci on developing a way to broadcast a message that will trick the alarm so that it doesn't go off when the landing party arrives and leaves."

"And this is feasible?"

"Theoretically. We won't know until we try it. I've been studying all of the aliens' intercepted communications going back to their challenges to war with the other species."

"Our mission launches in five hours. Get on it right away. Any other ideas?"

"We may not even need to breach the atmosphere," said Irv. All heads turned toward him. "None of those stations is protected by a force field. Perhaps our rivals don't have that technology. We could send a drone probe to one of the stations and that could do the trick."

The captain nodded. "I like your thinking." The object of the mission was to bring home a comprehensive database on the aliens. On most civilized human worlds such a total compendium of knowledge was commonly available on any and all computers, including the most common personal hand-held devices.

"Anyone else?"

The group was silent.

"Okay," said the captain. "Luci and Marji, get to work. Irv, launch that probe as quickly as possible. That is all."

Forty-five minutes later, Irv reported on the comm link. "Captain, the probe got partway to the nearest station then we lost contact. Likely a burst of solar wind fried it.

The captain let out a rare expletive, shocking the officer. "Send a second one."

"Aye, sir."

Moments later a small device jetted out of the cargo bay and sped toward a space station some 1,200 miles distant. Irv followed its progress on a 3-D schematic showing the humans' ship, the station and the space between. Once the device was close enough, the feedback switched to visual. The cloaked probe began to brake as it approached the alien station. "Closer, closer. That's it!" cried Irv as the probe attached itself to the skin of the station. A port appeared and a tarantula-like robot crawled out. Using a molecular scrambler, it pushed through the outer wall of the station. "Go, go you little monster," Irv cheered.

The gadget produced tiny wheels and it slid along various hallways, giving Irv a floor-eye view of the insides of the station. Because the gizmo was cloaked, none of the aliens detected it as it occasionally zigged around one set of legs or zagged around another. The machine stopped at a cabinet. The device produced an invisible arm. It planted a tiny hand onto to the cabinet. Moments later, Irv received a signal that the upload had succeeded. "Yes!" he said, punching the air with his fist. A short time later the bug exited the station, got back into its little automated vessel and zipped back to the mother ship.

Irv shared some of the data with the captain on a giant screen the security officer had produced with a wave of his hand. The captain used a finger flicking motion to page through some of the info. A number of pages were schematics, what seemed to be military info. But there was nothing of a more general, comprehensive nature. "Irv, this is good and headquarters should be thrilled with this data, but it's not enough. Apparently that station only keeps the info on hand that it needs to function. Maybe that's another security measure. Looks like we'll have to go planetside."

"Aye, sir."

The captain dismissed him and strode over to the ship's engineering section. He found Marji, three other engineers and Luci, each working intently with a wall-sized screen of equations and schematics. He approached the chief engineer, who had dark circles under her eyes. She was holding a steaming cup of *focus* in one hand. Several partially finished cups of *wakebrew* were scattered around the room. It seemed none of the star people were getting any sleep tonight.

"Ms. Faubner, how are we doing?"

She shot him a killing stare. "I'm working on it."

He balled his fists, pivoted and double-timed out the door. He hustled to the gym which, fortunately for everyone, was deserted. He began pounding the speed bag. The captain continued a vigorous workout routine. He checked the time. T-minus thirty minutes.

"Ms. Faubner . . ."

"Almost got it."

He started pacing the gym floor. If she couldn't soon figure out a way around the alarm it would destroy their time line, making them too likely to be caught up in the imminent war. And a still-functional alarm surrounding the planet would lead him to scuttle the mission and head out of this star system. There was no way he would expose his crew and the A'laamans to certain death. High risk on a mission was one thing. Being foolhardy was another.

These thoughts weighed so heavily on him that a burst of female voice startled him. "Captain, it's a go!"

"And you're sure this will work?"

"Ninety percent sure."

"That's better odds than we've had all day." He glanced at his wrist chronometer. T-minus eighteen minutes. "Landing party, prepare for launch," he said. He rushed through a shower and change of clothes. Hustling out of the gym, he made some last minute preparations but stopped in his tracks at a port that looked out on the planet below. The ship's orbit had taken them to part of the world where it was currently night. He saw darkened versions of deserts, mountain ranges, a couple of small seas. But no lights. None. *Where were all the cities?*

"Bridge, this is the captain," he called out louder than he intended.

"Aye, sir." said a voice on his comm link.

"I thought our sensors showed over a hundred-ten billion sentient beings on this planet."

"That is correct, sir."

"Then where *are* they?"

The crew member, looking out his port was equally dumbfounded. "I . . . don't know, sir."

No time for guessing. Departure was in seven minutes.

Sunrise was taking place on the part of the world the ship was over. Erik looked down at the planet once again. It looked orange, pock-marked and battered. He wondered how many previous interstellar battles it had seen, how many more remained. A quick adrenaline rush ran through him.

19

Somber Goodbye

ERIK SURVEYED THE OTHER members of his landing party: Montoya, Minj and Luci. All were wearing equipment packs. He had gone over the plans with them. This voyage had been full of surprises but the execution of the mission needed to be quick and surgical. In, get the prize then back out again, hopefully before the natives were any the wiser. They had gone over their plans and contingency plans. They had checked all the weapons and tools. The captain had spoken with Engineering. The lift shaft was back up and working so they should be able to get to the planet surface and back quickly.

A small knot of people stood nearby watching the final preparations. Erik's brow was knit and he held up his hand for silence as he listened to a message on his comm link. He shook his head then announced to the group: "Thousands of ships have entered the outer edges of this system. Assuming they get past the enemy's defenses, we expect them to be here in 0600 hours." Looking his flight director, Marji in the eye, he added: "If we're not back within 0500 hours, you have strict orders to leave this system using coordinates that will keep you as far as possible from the incoming ships."

She swallowed hard. "Aye, sir." she said, saluting.

He returned the gesture.

"Captain Houston, we wish you well," said the head of the A'laaman ship.

Daj walked over and hugged Lisa. She clung to him. When they separated, she was speechless.

"I love you," he said. They embraced again.

Some of those present shook hands with the various landing party members. Others slapped their shoulders or hugged them.

Erik saw Zama standing off to the side, staring at the floor. He walked over to her and noticed those brown orbs were wet. He opened his arms. This time she gave him a real hug and held him close. She whispered in his ear: "Come back safe, captain." They separated.

The landing party walked down a hallway and through a door that automatically closed after them. Zama followed several paces behind and stood in front of the door. Irv put a gentle hand on her shoulder. "Ma'am, you don't have security clearance," he said. "Besides, it's not safe for you to stand inside that door."

She nodded but continued staring at the exit.

A dozen yards away, the landing party had entered a huge, empty chamber and stepped inside a thirty-foot diameter circle that was inscribed on the metal floor. Erik produced a remote device from his pocket and maneuvered an ATV to the center of the circle. The group of four piled into the vehicle. Erik smiled when he recalled that the last time he'd been in an ATV . . .

"Initiating force field," said a voice in the comm link, snapping the captain back to the present. A force field bubble surrounded the ATV. "Commencing lift shaft," came the next message. Erik and the others saw the circle surrounded by bright light. The circle slid away to reveal the planet below, surrounded by the blackness of space. The ATV hovered in mid-air.

"Beginning descent sequence," said the comm link. The ship had created a temporary lift shaft that would quickly lower the vehicle to the planet surface. The shaft was like the ones used on inhabited planets to rapidly move people to different floors within a skyscraper or even to quickly transverse hundreds of miles from a planet's surface to orbiting space stations. To an outside observer, the ATV would appear to be rapidly descending within a tube that looked like a beam of light.

As the ATV entered the planet's atmosphere and continued descending the captain sighed with relief. There was no evidence of any alarms going off or any fighter vessels scrambling.

As the vehicle touched down, the shaft dissipated. The ATV and each crew member was cloaked to avoid detection by the planet's natives. The captain had given his officers strict orders to begin an evasive orbit so that even if the natives detected the recon party they would be unable to find the vessel based on the landing trajectory.

Erik was the first to set foot on the ground. His boot sank half an inch into the orange soil, sending dust billowing out. Montoya and the others followed. Weapons drawn, the visitors looked around in all directions. The red sun hung in the ocher sky. High overhead arched part of the planet's ring, disappearing over the horizon in either direction. Before them lay an extensive desert with a few scrubby trees and prickly brush. It was silent. Too silent.

The commander frowned. "Our instruments say we're on the outskirts of a major population center," he said.

Luci pulled out a hand-held instrument and re-checked it. "We're in the right place," she replied. "Our, uh . . . friends are a couple hundred feet below us."

"We'll have to find some unobtrusive way to get down there."

A tumbleweed rolled by. Nearby, small bits of debris hurtled through the air. Small whirlwinds here and there stirred up the sand. A distant whining broke out. The humans again looked in all directions. Montoya pointed to a low, orange-brown cloud that had started out tiny but was rapidly expanding.

Erik squinted his eyes. "Sandstorm!" he cried. "Everyone back in the vehicle! Force field on!"

Moments later, three people arrived back inside the ATV. "Where's Luci?" Erik cried. He stepped back outside. By now, a brown haze surrounded the area, almost blocking out the sun. The wind was blowing so hard it was he found it difficult to stand.

"Luci!" he yelled. "Luci!"

No response. The wind was now a loud whine. Streams of sand stung the captain's face and pelted his legs, his arms, his ears. He spat sand from his mouth. He tried to walk away from the vehicle to look for the missing officer but a gust knocked him to the ground. The vehicle rocked back and forth.

"Get in here or we'll lose you, too!" yelled Montoya. Daj and he strained to haul the captain back into the ATV. Montoya set the force field.

Erik sat on the floor of the vehicle. "Luci!" he cried.

Sometime later, the howling of the wind decreased then dropped off altogether. The brown fog began to lift. Montoya turned off the force field. The captain had to push hard to force open the ATV door. He crawled out onto a dune that had almost covered the vehicle. He was afraid that if he ever saw Luci again, it would be at the bottom of another dune.

20

Sandman Nightmares

AFTER SETTING FOOT ON the planet, Luci had walked about a twenty yards from the vehicle as she continued to check several instruments. Abruptly, it grew dark. She saw an outline of the nearby ATV but when she tried to head that way the wind blew her in the opposite direction. The hot surges of air knocked the diminutive lady to the ground then whipped her hair into her eyes, blinding her. She brushed the hair out of her eyes. She turned around and tried again for the ATV but the gale pushed her backward mid-step. She tried to set her personal force field but the instrument shot out of her hand, along with her cloaking device. The brutal wind reduced her to crawling on the ground. She spotted something ahead. A light was on in the middle of the storm. She dragged that direction, occasionally pushed off course by the powerful air stream. It wasn't just a light. *Something* out there was glowing.

As she drew closer, she noticed it was a short wall shaped like a triangle. She reached the wall and grasped it with one hand, then another. Forced by the weather to keep her head down, she became aware she was looking down into something. Was it a well? There was a series of wide steps on each side of the triangle, steps that narrowed as they descended. The three sides of the triangle met below, creating an inverted pyramid. She hurried down the steps, which were short enough she could take them two at a time. Was this some kind of storm shelter?

A cloud of sand blew into her face, making her feel like a hundred tiny needles had stabbed her. She struggled to keep her footing and coughed as breathing became difficult. Her hand found a railing and she gripped it. Eyes closed, she used the railing as a guide as she felt her way down the stairs. Within a few more steps, she had reached the bottom, the apex of the inside-out pyramid. She was only about a dozen feet below ground. Would this structure give her too little shelter after all? A ferocious wind howled overhead. She dared to look up. The sky overhead was reddish brown. She could see nothing more. Sand bit her face, her eyes, blinding her once again. She lowered her head and wiped her eyes. She crouched down near the apex. A patch of sand poured down the steps and settled at her feet. Then more arrived. Up above, a dune that had formed near the walled structure collapsed, sending more choking dust and sand cascading into the pit. Even more fell from the sky. The tiny steps were becoming sand waterfalls. The sand flow buried her ankles. The sand was too heavy for her to move her feet. The invading substance began to reach the crouching woman's legs. Soon it was at her hips, her waist. Luci was coughing as she tied a neck kerchief over her mouth and nose. She could feel the grit in her ears, she crunched sand with her teeth. Her breathing became more labored. Her hacking sent sharp pains through her chest.

Her mind flashed back to the nightmares she had been having on the trip to this planet. In those dreams, she had been out of breath. Her heart was pounding as she ran full speed. She was in a vast, dimly lit cavern. A frightening creature, perhaps several were chasing her and she felt unable to outrun them. Sharp, jagged sensations kept nipping at her arms, her legs. She tumbled to the ground and beings descended on her and began picking at her, tearing at her. She covered her face and head with her arms. She screamed.

Startled awake, she would sit up on her hover field, trembling. She had been having the dreams nightly ever since the ship had gone into space. This had concerned her because she rarely even remembered her dreams. But each time the dream occurred she was afraid to go back to sleep.

After one especially tough night she had arrived the next morning at her shipboard post, her right hand hugging a cup of steaming *wakebrew.*

Marji had noticed the dark circles under her friend's eyes. "What's wrong?" she had asked.

Luci had pulled her aside in order to be out of the captain's hearing range and had said in a soft voice: "Don't tell anyone else, because I don't want to hurt morale, but I've got a bad feeling about this mission."

When she had gone to sleep that night, she had dreamed she died.

Her mind coming back to the present, she choked on the hot sand and wondered if the death part was imminent.

"So this is how it will end. Being buried alive on an alien world," she thought.

She passed out.

21

Descent into Hell

L UCI'S FIRST SENSATION WAS being bounced and jarred as if she were on a trampoline. She felt sand in her hair, in her mouth, all through her clothes, on every square inch of skin. Her face felt as if it had been attacked by sandpaper. But the wind was no longer blowing and no more sand was pelting her. It was no longer oppressively hot. She heard a distant humming. *Was it some sort of machinery?* She opened one eye a sliver. What she saw made her gulp a breath so quickly that she almost went into a choking spell. Four insect-like creatures, each about three-quarters of her height, were carrying her, each one holding a limb. Each being walked on two legs but had four arms that ended in fuzzy feelers. The creatures had almond-shaped heads and yellow eyes. They didn't smell good, either. She wrinkled her nose at the pungent stench. She wondered if she were hallucinating but this seemed all too real. Squinting, she noted they were in a vast, dome-ceiling expanse that was well-lit but reminded her of a cavern. Unlike a natural cavern, there were no stalactites or stalagmites. Her stomach felt queasy. This setting reminded her too much of the nightmares.

The natives stopped at a round dias and dumped her onto it. A small black snake wriggled onto her and she recoiled. One of her captors grabbed the viper and threw it. The being pointed at the reptile, which burst into flames. The insect people gathered around Luci and stared at her as if she were a specimen. A tiny, round mouth appeared on one of them and a pink tongue as long as a rope uncurled, touching Luci's arm, giving her a wet, sticky sensation. Next, one of the alien's large feelers stroked her arm. The sensation was like running a huge, soft-bristled brush over her skin. She shivered with revulsion. Her adrenaline running full blast, she grabbed the tongue with both arms and yanked on it with all her might. The offender doubled up and the others pulled back, tearing the feeler off her skin as well. Her arm felt as if a giant band-aid had been yanked off. The lady cringed for a moment. She rolled off the dias and sprang to her feet. She ran off, knowing she was about to die but she ran anyway.

Her longer legs quickly put some distance between her and the bug people.

Her pursuers shot ribbons of flame after her. One of the fire bolts singed her arm. She gritted her teeth and kept running despite the pain. As she poured on the speed, her eyes shifted to every section of the cavernous room. The place seemed to be on such a large scale that she couldn't tell what was seeing. Then: salvation ahead! A shaft of sunlight pierced the dimmer light of the cavern. She spied a tiny ladder than ran up a wall. It was difficult to judge its distance. Fifty yards? She hoped no further. She glanced behind her and hoped these weren't flying bugs. Her legs pounded harder.

The ladder was starting to grow in size. She heard some loud clicking sounds. She ignored them and kept running. Her goal was in sight. She could almost touch it. She leaped onto the ladder and all four limbs went into overdrive. Her hand-over-hand motions were working as fast as her legs. Luci was gleeful. She felt like that mythological creature, the monki. She glanced up. The circle of light up above was growing larger. She ignored the soreness of her limbs and kept pounding away. Ground level appeared to be forty feet away. Thirty. Twenty.

Then a fuzzy feeler cupped her mouth. Now one, two, three, four slimy rope tongues wrapped around her limbs. Pried off the ladder, she felt her body descending. He heart sank through her chest.

She struggled, not caring if it shook them all off the ladder. It would be better to fall to her death than to be at the mercy of these creatures.

She felt her stomach drop as the aliens scaled down the ladder more quickly than she felt possible, even faster than her ascent. Upon reaching the floor, they hustled her to another dias, this one oval, and threw her down. Her shoulders, lower back and buttocks slammed into the hard surface, jarring her.

"If you're going to kill me, make it quick," she thought.

"*You don't get off that easy,*" said a raspy voice in her head.

They could invade her mind! She suddenly felt naked.

"That's right! You'd better be afraid," another alien voice rang.

She swallowed hard.

The creatures looked at one another and made some rapid clicking sounds but she could also hear faint mental voices.

"Just immobilize it and remove its brain. That will give us what we need," said one.

"Too much work," said another. "We'll just terrorize it into cooperating."

"We should let it go. It hasn't harmed us."

"That's right. We've got bigger things to do."

"I give the orders here. This thing is a threat. It's keeping vital secrets."

"We're in over our heads. We need to turn this . . . creature over to our superiors."

"Forget that! I want to get the credit for getting the info."

"Enough! We're wasting time!"

The voices in her brain ceased as her captors seemed to revert to just the clicks and gestures for communication.

One of the insects flexed its elbow, revealing an inches-long stinger. Luci gulped. One of his comrades placed a restraining feeler on his companion's arm. The two bugs locked eyes. The creatures exchanged some rapid clicks. The insect with the stinger pointed away from the area. The other lowered its head and slunk away. The offending bug flexed the stinger and jammed it toward Luci. She tried to roll out of the way but the remaining two creatures held her still. A sharp pain jabbed her arm, followed by an intense burning. The attacker yanked the stinger back out. She yelped as she felt her flesh tearing. Several deep lacerations turned red and small streams of blood ran down her arm. Within seconds, she felt waves of a warm sensation radiating outward from her arm. She was unable to move any of her limbs.

The attacker stuck its face within inches of the human's. She choked on the evil stench of its breath. She saw tiny images of herself in those faceted eyes. Her breathing was labored. Her heart was pounding out her chest.

"*What are you?*" hissed a voice in her mind. "*Why are you here?*"

Trembling, she struggled to control her thoughts. *"Can't think about the mission."* She closed her eyes. She remembered the party clothes she'd worn to one of her childhood birthdays. The neighbor kids and her cousins were playing games . . .

Her memories projected on to a far wall of the vast chamber. Here and there, knots of the ant people stopped to gaze at this incomprehensible sight.

Her thoughts shifted and now she was on her home planet. The red sun and its white dwarf companion were in separate parts of the sky. She was now an adult and sat on a blanket under a tree while a guy and she enjoyed a picnic.

"This data is useless!" shouted the alien voice in her mind. *"Why are you here? Why now? Give me what I need!"*

"Can't think about . . ." Luci gritted her teeth. She bit down until her tongue drew blood. Numbers began to appear on the far wall. Some were in color. Some were plaid.

"You're a spy!" cried the voice in her mind. *"You didn't come all these light years by yourself. Where are the others? Where is your ship?"*

She kept trying to block out the creature's demands but she felt control of her mind slipping.

"Where is your ship?"

"Be strong," said her mind. *"Be strong."* Her lips mouthed the words.

"Enough!" shouted the mental voice. "Your species is weak. Once we get through our . . . present circumstances, we're come back and exterminate you, one planet at a time. You'll be unable to resist. Just like you'll be unable to resist me now."

The bug leader stuck its head on her face. Its head was covered with a powdery substance that made her skin feel on fire. She closed her eyes again. Tears ran down her face. She remembered her training from years ago about how to avoid giving out any vital information even under torture. She thought back to the picnic. And the guy.

The creature cupped her chin and lifted her head, yanking her back to reality. "Do others of your kind consider you beautiful?" it hissed, pausing to wait for a response "I thought so. We'll destroy your beauty." An antenna on its head flexed then snapped like a whip, cracking in the air.

"Last chance," hissed the voice.

A pause. The antenna snapped again, tearing across Luci's face.

She screamed. A long stream of blood formed along the laceration.

"Your shell is soft," the voice in her mind said. *"You won't last long."*

The ant man backed off and began to grip and ungrip the jagged face pincers that bordered its mouth. Another creature moved its face menacingly close to Luci's arm.

"You're out of time," warned the chief tormentor.

Hell began.

22

The Jaws of the Ant

MINJ WAS STANDING ATOP the dune that had almost engulfed the ATV. "I've picked up Luci's bio chip," he cried, glancing at his hand held device.

The captain breathed a prayer of thanks. "She's alive! Help me clear away this sand." he snapped.

Minj shook his head. "She's not there. She's deep underground."

Montoya's jaw dropped.

"What's the quickest way there?" asked the captain.

All three men scanned the area for an entry point to the underground world. Montoya searched through the power binocs. He pointed to a short tower in the distance. "Looks like a ventilation shaft."

Erik and his crew set their antigrav belts and launched into the sky. "Set your cloaks," he said.

The star men arrived at the shaft, eased through a wide vent and descended. They emerged in an enormous, cavern-like chamber. Several elevated roadways and bridges criss-crossed the open space. Combinations of vehicular and foot traffic traversed these conduits.

Alighting on the floor, the humans each aimed a scanning device at the walls of the vast cave. The scan revealed that the walls were not solid rock but were honeycombed with numerous small chambers each occupied by a group of natives. The caverns were apparently public areas. Schematics indicated there were lower levels beneath this one and even deeper ones going down for miles.

"She's this way, " said Minj, pointing. They took to the air again and flew some distance. The group of humans had set up a sound block so the aliens could not detect their speech.

They flew through a vaulted hallway that led to another over-sized cavern. Numerous individual aliens and long columns of them moved along the wide hall. The cavern led to tunnels that branched off in several directions. Minj held up his hand, causing the trio to stop.

"What's wrong?" asked Erik.

"I've lost the signal."

"B-but the chip will transmit as long she's alive, right?"

"Unless something's blocking it."

Montoya studied his hand-held. "I've got her." He frowned. "Vitals are off the chart. She's suffering severe trauma. This way! Hurry!"

They sped through a long, vaulted hallway. A high-pitched female scream pierced the air. The trio increased they speed. The captain went into full battle mode at the sight that met him. Luci lay on an oval table, bleeding profusely from her face and arm. A large chunk of each was missing.

"You savages are dead!" shouted Houston.

He pulled out an ener blaster and laid waste to the three creatures surrounding the injured crew member. The men lighted on the ground. They turned off their cloaks so their crew mate could see them. Her eyes were glazed and her skin was white. The group re-cloaked. The dias became a sick bay as Montoya pulled off his med pack and sprayed some *bleedstop* on the gaping wounds. The blood coagulated. The hapless Luci tested positive for infection. The doc placed some *quickabsorb antibiotic* gel on her tongue. He treated her for shock.

The captain noticed a 3-D video on the fall wall of the chamber. It was a shifting melange of color with images of humans dancing. "Shut that thing off!" he snapped. "We'll have their whole city over here!"

Minj did an instrument scan. One of the bugs had tapped into Luci's brain. Minj kicked one of the fallen creatures in the head. The wall video flickered and disappeared.

"I can't move her until she stabilizes," Montoya said. "The cloaking should buy us some time. Can you two handle the mission on your own?"

The captain nodded.

"Let's go D-base hunting," said Minj.

The doc waved a diagnostic over the wounded crew mate. "Vitals are improving," he reported.

"Take care of her," said the captain.

The doc nodded.

The captain nodded to Minj and the two launched back into the air.

Montoya held his patient's cold hand. "Hang in there." he said. "We'll get you out of this."

Minj followed his instruments through a long passage and into another vast maxi-chamber. The flying duo, still undetected, eased to the ground.

"Where's that computer?" asked that captain."

Minj pointed to a solid rock wall in front of them. "I can't find a way in," he said.

Houston retrieved a hot plasma torch and was about to engage it when Minj laid a restraining hand on his arm. "Look at this."

Erik and he studied the display on his hand-held device that showed the scene behind that rock wall. A knot of locals was gathered around a large piece of machinery. A 3-D projection of the local solar system filled the immediate area, showing a myriad of dots approaching from one direction and another grouping coming in from a different way. The dots were 'way too close to the ringed planet.

"The two navies are almost here!" cried the captain. "We've got even less time than we thought."

The men moved about a dozen yards along the wall so that when they created a way into the chamber the machine would block the aliens' view, preventing them from realizing anything was awry.

"Any alarms on that chamber?" asked he captain.

The other man nodded. "A couple simple ones. I'll have 'em neutralized right . . . now."

On Minj's signal, Houston activated the plasma torch. The device melted through the rock as he carved out a large circle. The process took a few minutes because the rock was a couple feet thick. The two men kicked repeatedly at the circle until the section of rock eased inside the wall, making a crawl space large enough for the diminutive Minj. He hurried through it and was now behind the huge machine.

"This system contains a library-sized database," he said. "Should give us what we need." He started punching keys on a hand-held as he tried to hack into the alien system. Half a minute passed.

One minute. Two minutes. He shook his head. "Their firewalls are good," he said. "I just hope I can get in without alerting them." His fingers again flew over the keyboard.

Three minutes passed. Five. The captain was praying for Luci, praying their mission would somehow succeed, that none of them, would die including the A'laamans.

"I'm in," said Minj.

Erik punched the air with both fists.

The data geek clicked a button to upload the megafiles. After a few seconds, the hand-held confirmed that the upload was complete. He ran a couple quick diagnostics to make certain the info was in a usable format. "Paydirt!"

"Let's get outta here," said the captain.

The 3-D sphere showing the solar system and advancing navies flickered and disappeared. The knot of locals turned toward one another and clicked loudly.

"Not good," said the leader. "First they capture Luci then we knock out their tracking system. If they figure out how to penetrate our cloaking, we're goners."

Erik and Minj retraced their flight path and within a few minutes were approaching the site where they'd left Montoya and Luci. They looked down on a battle scene of Fred, force field in place, firing valiantly to hold off a half dozen attacking insect people. Erik and Minj joined the fight from above, quickly dispatching the attackers. They eased to the ground beside their comrades.

"We've got to get her out of here. Now," said the captain.

"Since she's my patient, let me carry her," Montoya replied. He looked down at her. "I hate to do this," he sighed. He put one arm under her shoulders, another under her legs and lifted.

The humans all took to the air just a small swarm of locals arrived. The captain and Minj kept them busy with their weapons. As they turned their cloaks back on and sped back toward the vent shaft, additional groups of natives began firing at them.

"How are they spotting us?" cried Erik. "Are our cloaks still working?"

Minj checked his hand-held. "The cloaks are fine. They may be following our trajectories from when we turned off the cloaks."

"Evasive maneuvers," ordered the captain." The group veered off in a different direction. The attacks continued.

"They may be tracking us based on our brain activity," Montoya speculated. "They were able to read Luci's mind."

"Maybe that doesn't work from a distance but just in case, do not think about the ship or its location," said Erik. "Okay, guys. Shortest route out of here!"

A group of aliens set up a large weapon and shot, hitting the starmen's force fields with enough power that they all felt the impact. Montoya, jolted, almost dropped Luci. They took another hit. But they were unharmed and were able to keep advancing. They reached the shaft. They ascended at full speed.

The humans zipped out of the vent and into open air. Within moments, Houston pointed at the ATV below. They alighted next to the vehicle. Montoya eased Luci to the ground and injected some additional meds into her arm.

"How is she?" asked Minj.

"Still fighting," the doc replied.

The captain officially broke comm silence with a call to the ship. "Mission completed," he said. "Get us outta here."

"Aye, sir." came a voice in his ear.

Long seconds passed. Nothing happened.

"Get us off this rock. That's an order.

Another long silence. "Sir . . . the lift shaft mechanism *isn't working*."

The captain picked up a rock and whipped it at the ground.

Fifty yards away, a long fissure opened in the earth and dozens of aliens began pouring onto the desert floor.

"Weapons, full power!" snapped the captain. The three men bathed the fissure and the surrounding area in fire. The aliens dropped and began writhing on the ground. A second wave emerged, then a third. The natives began firing. The starmen set their force fields but the group spread out and surrounded them.

"Captain, we'll fire on your command," the voice of Irv Malvo said into his ear.

"Negative. You'll give away our ship's position."

"But sir, how long can you survive without backup?"

Houston's reply was breaking up as a heavy barrage hit his force field.

23

Help

ZAMA WAS CHEWING HER fingernails as she watched the monitors. Given the importance of the battle Irv had relaxed the rules about only authorized personnel being allowed on the bridge. So Zama had seen Erik's team pinned down under heavy fire and had witnessed a tense Malvo dispatch a landing pod carrying a handful of backup fighters to the planet's surface. The reinforcements had landed on the other side of the enemy, which had begun taking fire from two sides. The new warriors dropped several of the aliens. Then a small tower arose from the ground and fired on the now-unoccupied pod, disintegrating it. Zama slammed her palm onto a counter, startling a few nearby observers.

As the fighting continued she watched first one then another of the backup troops take hits from enemy fire. Neither person got back up.

She paced the floor and bit her lower lip. When it got too intense for her to watch, she strode over to one of the ports. She gazed at the stars to try calming her nerves. She saw several tiny flashes of light in the distance. Was this the fleet belonging to the other species, now in another part of the solar system and dealing with some of the same hazards that had afflicted the humans a day earlier? She realized that light from that distance would have taken several hours to reach their ship so the invaders were much closer than they appeared. *Too* close. She strode over to the leader of the destroyed A'laaman ship.

"Captain Mies, we've got to *do* something," she said.

He nodded somberly. "What *can* we do?"

"That life boat of ours, it can also serve as a landing craft, right?"

"Yes."

"Check to see if it's still working."

"Yes, Ma'am."

As the captain hurried away, she turned toward the assembled crowd and called out: "How many of you helped lead our troops in the war with the Batu?" Most of the A'laamans raised their hands. She pointed to a man she recognized as a decorated officer from those years-ago battles on A'laama. "Have you kept up your weapons skills?" she asked.

"Back home I hunted three or four times a year."

"Good enough. You're going to lead us."

He swallowed. "Yes, Ma'am."

She pulled an ener pistol from her pocket, weighing the silvery firearm in her hand. It had been in a hidden compartment of her overnight bag. The weapon had belonged to her late husband. She had never fired it prior to her trip into the wilderness with Erik when they had encountered the hostile Batu tribe. In the years since, she had become a hunter and an expert marksman. She glanced at the gun and frowned. "This pea shooter won't be enough," she said. "Does our life boat have any ener rifles?"

"Yes, four or five of them," the newly appointed leader replied.

"Somebody get me one. You have five minutes to teach me how to use it."

Irv has been standing out of earshot. She hurried over to him. "Mr. Malvo, how much time until the fireworks start?"

"At least one of the navies should arrive in about three hours and fifteen minutes," he said.

She turned back to her battle leader. "Correction, you have two minutes to teach me how to shoot it."

"Yes, Madam Exec."

"Wait a minute," said Irv, placing a restraining hand on her shoulder. "I can't let you go planetside. You'll get yourselves killed."

"But if we don't, more of your people may die."

Irv held his ground. The two of them locked eyes. He looked back at the monitor for a long moment. He slowly nodded. "Do you realize you only have a little over an hour to get the job done, then our ship will have to leave, no matter what?"

'We'll take that risk."

The other A'laamans nodded.

"Then fight well," said Malvo.

"I want those aliens to feel our pain," said the newly appointed A'laaman leader.

The impromptu troops filed out of the room and the officer gave Zama a quick lesson on the ener rifle. The A'laamans assembled in the launch bay and held hands while the former chiefexec said a quick prayer. The cavalry piled into their vehicle. The cargo bay door opened. The A'laaman landing craft leaped out into space. As her transport hurtled through the atmosphere toward the orange surface below, Zama steeled her nerves. The starmen could not die. They *could not* die.

The vessel shot out of a cloud and toward the desert. A miniature battle scene rushed up to meet her. She shook her head. It looked grimmer than she had imagined. While still descending, the craft fired heavily on the aliens, thinning their ranks. A new fissure opened in the ground and an army poured out. The A'laamans concentrated their fire. They then set up a force field over one of the fissures, sealing any additional aliens inside. The tower shot a beam of light at the A'laaman vehicle. Its force field automatically kicked on, averting catastrophe. Zama began hyperventilating. Severe pain shot through her chest and arm. Maybe they would all die. But not without a fight.

Her vehicle touched down. Another fissure opened. The A'laamans sealed it, stopping the ground from hemorrhaging any more of the enemy. The leader commanded his troops out the hatch and onto the desert floor. Zama staggered out of the craft, still hurting. *"C'mon meds, where are you?"* she thought.

The A'laamans had landed near the second wave of star men, with the aliens in between that group and Erik's team. Zama froze as she spotted Erik about fifty yards away. As he noticed her, his face turned white and his jaw dropped. She stopped shooting for just a moment and stuck her thumb high in the air. He slowly nodded.

The alien tower shot another beam at the A'laaman landing craft, disintegrating it. The tower abruptly melted. Zama assumed the star men had concentrated their firepower on the weapon.

She returned the favor, promptly dispatching first one then two aliens that were bearing down on them. She saw a creature produce a long rod that appeared to be made of light. An explosion knocked her back despite her force field. She recovered her balance. A second weapon hit her from behind. Instantly, her shoulders felt on fire. She fell to the ground. "Help us, Lord!" she moaned. She blacked out.

24

Cutting the Losses

B ACK ABOARD SHIP, IRV double-timed it to Engineering. "Lieutenant, we're out of time," he said.

"Sir?"

"You heard the captain's strict orders. We've got to leave the premises an hour before the alien navies are expected to arrive. That's exactly an hour from now. We need to start preparing to leave this system."

Marji stepped away from some diagnostic equipment, turned her head around and looked at the acting commander. She swallowed hard. "Sir, I'll have my senior engineers handle the departure prep. I've got to keep working on this shaft generator or we'll never retrieve everybody."

Irv's face was pale. "W-we may not have time. I hope it doesn't come down to this but we may have to sacrifice some so that others can live."

The head engineer grew quiet. Luci's face flashed through her mind. She thought back to the deep discussions the two of them had had when she'd fallen in and out of love with Jev, then back in again. She had also given Luci moral support through several different relationships.

She thought of the years she had worked with the captain, the times Montoya had treated her for Sildovoran flu and various aches and pains. Even her years-ago rivalry with Minj now seemed like a pleasant memory.

"Lt. Faubner. Marji. Are you getting this?"

She nodded. "I understand, sir, " she whispered.

She returned her attention to the lift generator. While she was immersed in her work, the minutes sped by. She barely noticed a large hand on her shoulder until it tightened its grip. Her heart was in her throat as she looked up at Irv's grim face.

An automated female voice broke the silence. "Departure in ten minutes. All officers and crew: make final preparations. Departure in nine minutes and fifty seconds. Repeat . . ."

"That means you, lieutenant."

She yanked on a strand of hair, oblivious to the pain. Her eyes bore into the superior officers'. "Five minutes."

"Three."

Beads of perspiration mingled with a tear that rolled down her face.

"Departure in nine minutes and thirty seconds," said the mech voice.

25

Game Changer

Rik's heart sank as he saw Zama fall. He pulled out his mega binocs. She still appeared to be breathing. "Fred, any way you can get to her?"

"If we both stop firing, we'll get all us all killed," Montoya grunted. Minj growled in agreement.

The captain switched on his anti grav belt.

"What are you doing?" snapped the doc. 'If you fly over to her, every alien around will open fire on you."

"Shut up, Fred," yelled the captain, who burst into running. He hurried across the no man's land between the two contingents of humans, coming within twenty yards of the alien troops. He arrived at Zama's side. She was lying face down, several layers of skin in her shoulder area having been burned away, as was the part of her shirt that covered her shoulders. Her shoulders were deep red. They slowly rose and fell. He glanced over at the fallen A'laaman leader. He was no longer breathing.

The star man returned his attention to Zama. He took off his antigrav belt, wrapped it around her waist and twisted some knobs. "I'm sorry if this hurts," he said. He eased down on his haunches, slipped his left arm under her head, his right arm under her legs and slowly stood up with her. The antigrav belt helped him lift and balance the lady, who was a few inches taller than he. "You're going to be all right," he whispered in her ear. "We're going to get you out of this." As quickly as he was able, he moved back toward Montoya, Minj and the prone Luci, who had been lying on the ground throughout the battle. He laid Zama down close to his fallen officer and went back to firing at the hostiles.

"How much time?" Erik comm linked to Irv.

"0210 hours, sir."

"How's that lift generator, Lt. Faubner?"

'Still working on it."

Houston went back to fighting. He located the alien who had shot Zama and fired on him repeatedly until he dropped. The wounded lady stirred and moaned. Erik fought back a tear.

Several aliens launched toward the small knot of star men, continuing to fire. One of them ran upon Erik but bounced off his force field. Were they attempting hand-to-hand combat?

Fifty yards from the humans, yet another crevice opened, sending dozens of new defenders into the fray. They began firing on the invaders as well. Multiple blasts hit the star men's force fields. The barrage continued for several minutes as the star men fired back, dropping several more natives. Erik saw the iridescent glow of his shield flicker than disappear. He rapid repeat fired into the crowd of attackers, gritting his teeth. Moments later, the shield came back on.

The locals continued their fierce fighting. "Keep firing," Erik called out to the remaining troops. "We've got to protect Zama and Luci!"

"Sir, we have a lift shaft!" cried a voice in his ear.

"Freeze departure sequence," said an older male voice in the background.

"Departure sequence halted," said the mech voice. "Countdown stopped at T-minus two minutes fifteen seconds, repeat . . ."

The captain swallowed hard. "Engage!" he cried.

Before the shaft began forming, the force field surrounding Luci faded and died. Several of the enemy rushed forward and picked up her body, scurrying off toward the crevice. Erik looked his second in command in the eye. "Fred, take care of Zama," he said. "I'm going after Luci."

The doctor looked down at Zama, who laid on the ground in a crumpled heap, then looked back up at his commander. "I will," he said.

The captain turned and hurried after the kidnappers. Minj followed quickly behind.

"But captain, you'll miss . . ." Fred called out.

Within seconds, the shaft began surrounding the remaining star men.

26

Elevator into Space

As the lift shaft an formed an outside observer would see what looked like a beam of light include a radius that surrounded the star men, Zama and several of the aliens.

"Can't we leave the enemy behind?" called Montoya.

But it was too late. The shaft had cut some aliens in two, killing them instantly. Portions of aliens stayed behind on the planet while the shaft drew other pieces along with it. Those who had gotten within the proscribed thirty foot diameter remained whole and were being lifted toward the star men's ship, too. While the humans stayed still and waited as the shaft drew them higher into the atmosphere and toward space, the aliens thrashed violently about. The group was soon back in the ship, bug people and all. As the shaft deposited its living cargo on the inscribed circle in the cargo bay, a force field surrounded them until the bay was re-pressurized so the humans could breathe without assistance. As the force field dissipated, the aliens became re-oriented and returned to battle with the humans. One of them saw the prone Zama and rushed over to her, raising its elbow talon into the air and preparing to jam it into her. Fred ran and tackled the diminutive offender, knocking it to the ground. He smashed his fist into the creature's face, knocking off one of its mouth pincers, which continued to open and close on the floor. A crew man fired his ener blaster at the insect, finishing it off.

Another alien aimed a rod of light. Fred pictured the weapon blowing a hole in the ship. He waved his hand, creating a force field around the combatant just as it pulled the trigger. The resultant blast, contained by the force field, consumed the alien in a burst of flame.

The door to the main body of the ship burst open and four armed crew members charged in. They aimed and fired, wiping out the remaining aliens. "Get them in *cryo bags,* one per bag," Montoya ordered. Headquarters had provided a number of the bags as a means to preserve any alien corpses until they could be studied at home. Following behind the armed crewmen were med techs who used an anti-grav stretcher to transport Zama to the infirmary.

"Okay, everybody. Back into the ship," Montoya said. He called Marji on his comm link. "As soon as this area is clear, create another shaft so we can get the rest of the humans out of there! Try not to snag any aliens this time."

"But I can't control . . . yes, sir."

She reduced the shaft diameter to fifteen feet and initiated the shaft. It retrieved all the remaining humans and fortunately, no bug people. The med techs returned to carry the two wounded star people and one A'laaman to sick bay. Several crew members who had perished in battle, as had the leader of the A'laaman troops.

As the exhausted Fred dragged back to the bridge, a crewman handed him a food bar and a cold drink. He nodded his thanks.

"Sir, two space stations have just launched all their fighters!" shouted a crew man.

Montoya's heart jumped. Within moments, space lit up with a number of nearly simultaneous explosions, destroying five alien vessels. The remaining ones fired back at their attackers. Additional stations sent more of the natives into combat. The human ship's instruments registered a fleet several thousand vessels strong—the defender's navy. Arriving from another direction were their enemies.

Several flashes were occurring on the planet's surface. "Lt. Montoya, the *other* aliens must have picked up reinforcements," said Irv. "They've got almost three times the number of ships as our hosts."

"Poetic justice," said Fred.

"Aye, sir."

Beams of light shot up from the planet's surface. A myriad of additional defenders launched from the planet into space. As the fireworks intensified, the human's ship began to rock.

Trembling inside, the lieutenant realized it would be impossible to locate and rescue Erik and the others without endangering everyone aboard ship. The only option was to follow standing orders. The orders to abandon longtime friends to save the rest of the crew and the A'laamans. And the star men didn't even have have their prize, the reason they had come all this distance and braved so many obstacles. The data disk on the aliens' civilization was back on the planet with Minj. His knees buckling, Montoya eased into the over-sized captain's chair. "Flight crew," he said, "get us out of here!"

"Yes, sir."

The humans restarted the automated countdown and made final prep to break free from orbit and flee the battle. Now all they had to do was go back through the aliens' obstacle course from hell.

27

Setback

ZAMA WAS ALONE IN a dark room. She was partially awake and lying on her side, the only position that gave her some relief from those flaming shoulders. Her hair was in disarray. Her chest slowly rose and fell with her breathing. Her head throbbed. Her foggy mind was trying to make sense of her surroundings. The last thing she remembered was those ugly bug people and a blinding flash of light. Now, she was somewhere else. It didn't feel like heaven. Was she a captive?

She heard a sound that was getting closer. Footsteps? She didn't think the aliens sounded like that. She opened her eyes a sliver. It was the star man doc. Had they somehow made it back to the ship?

He waved his hand and the room was suddenly a little lighter but not so light that the change blinded her.

"How are you feeling?" he asked, sounding far away.

"A little better," she moaned in a deep voice.

"Let me take a look." A small blanket hovered a few inches above her shoulders, shielding from view the area where the shoulder-less cloth gown ended. He walked behind her and gently peeled back blanket, revealing a sea of large blisters. "You're making some progress," he said. "How is the pain?"

"It's…okay."

"I didn't ask if you could endure it. Do you want some help?"

She stiffened and gritted her teeth. "I'm fine."

"I hate to see you suffer. Are you awake enough to take a pill?"

She nodded.

He reached into his coat pocket, pulled out a vial and opened the cap, shaking a tablet into his hand. "Here."

"Thanks, doc." She opened her mouth and he placed the med on her tongue. She grasped a nearby cup and took a gulp of water.

"You know, I thought you crazy fools would get us all killed," he said.

She bit her lip.

He continued: "But . . . that was some strong fighting."

Her lips curled into a smile. "You, too."

He smiled as well, one of the rare times she had seen him do so.

Silence reigned. She tried to look him in the eye but he avoided her gaze. "Dr. Montoya . . ."

"Yes, Ma'am?"

"Where's Erik? He did . . . make it back okay, right?"

The physician looked at the floor.

She reached out and gripped his arm. "H-he is still alive, right? Tell me he's still alive."

He let out a long, slow sigh. "I suppose there's still hope. He kept telling me to have faith."

Her eyes grew wide. "W-what are you talking about? Where is he? Where's Erik?"

"I-I'm sorry, Ma'am," said he said, hanging his head. He turned and shuffled out of the room.

A tear rolled down her face. Despite the meds it was a long time before she was able to go back to sleep.

28

Surrounded

ERIK TORE AFTER THE aliens who were carrying Luci. He stopped, aimed and fired at a bug creature who held one of the crew member's legs. The being collapsed, letting go of his captive. That brought the group of captors to a halt.

Minj arrived at the captain's side, panting from the run. "Look," he said, pointing to a column of light in the distance. "We've missed the shaft."

"We'll try to catch the second run," the other man snapped, rushing forward and firing at the other three captors. Reinforcements emerged from the nearby crevice and helped carry the injured human. The group turned around and headed back underground.

The captain swore as Minj and he resumed pursuit. They were soon below the surface, hustling down a long ramp that ended in one of the giant, lighted caverns. Still more aliens joined their companions. They began firing at the would-be rescuers. Multiple blasts were hitting each man's force field. Minj's flickered then went out. Erik's shield, also overwhelmed, terminated.

The captain swallowed hard.

The star men fired back at their attackers.

Several aliens, weapons drawn, encircled the two humans. One of them stepped within a foot of Erik.

"So . . . you're the leader," hissed a voice in his mind. "You've gone to a lot of trouble to protect that worthless piece of meat." The mind invader spat a green fluid toward Luci.

The captain's muscles tensed. He fingered the trigger of the ener blaster.

"Oh, I wouldn't do that. Otherwise we'll revive your friend and make certain her last moments are the most tormented of her life."

Erik was about to fry as many of the bugs as possible but in a split second the closest alien jammed its elbow talon into his thigh. The human's leg suddenly felt on fire. He collapsed to the floor.

Minj continued to hold his ground, keeping his weapon ready and sweeping his aim from one alien to another. The antagonists ignored him.

"So, how shall we proceed," said the mental voice. "I can either make you watch while I dispatch this pitiful specimen," (it tilted its head toward Luci) "or, I can have you endure the same experience." It moved its mouth pincers and snapped one of its antennae like a whip.

"Quit playing around," said another mental voice. "We don't have time for this. Just finish off these vermin and be done with it."

"You don't understand," the captain thought. "We don't have to be enemies. There's a lot we can learn from another. We can help one another."

"We know all we want to know about your kind."

"Enough already!" The second creature pointed an arm toward Erik.

Luci opened first one eye then the other. She reached into a pocket and pulled out an ener blaster. She aimed and took out the creature that was threatening Erik. She began firing at others. Daj did the same.

Houston leaned up on one elbow, aimed and fired. An ant person aimed his arm at the captain. He felt a small blast then blacked out.

29

Life Line

ERIK FELT GROGGY. THE lower half of his body felt numb and the other half was sore in various places. He opened his eyes but it took them a moment to focus. He spotted Luci about ten feet away, sprawled out on the ground, once again unconscious. Her gun was at her side. Her skin was white but her chest was slowly rising and falling. He sighed with relief. He turned his head and saw Minj several feet away.

"Buddy, you alright?" asked the captain.

"Yeah. A little sore but okay."

Erik struggled to his feet. His left leg, still somewhat numb almost gave way under him. No aliens were nearby, at least not any that were still alive.

Minj pulled up into a standing position and brushed dirt off his clothes.

"We're not cloaked, yet they've leaving us alone," the other man mused.

"Spooky, huh?"

They looked around. Red lights were flashing at several distant locations and large groups of aliens rushed about in various directions.

"Maybe their company has arrived," the captain mused.

"Could be our chance to escape," said Minj.

"Escape . . . where? Our ship would have already left. We're stuck on this planet."

Daj pulled a hand-held device from his pocket and inserted the disk containing the alien data base.

"What're you doing?" asked Erik.

"A search. Ah . . . look at this." He showed the captain a schematic.

"I'll be. How far away is this?"

"Two point eight miles."

"First, let's make sure all our equipment is functioning." They checked out their invisibility cloaks, the force fields, the anti-grav belts. Everything was in working order.

"Okay. Let's hurry before our . . . friends get there," Erik said.

Minj eased onto his haunches and picked up Luci. The two men set their cloaks and shields then launched into the air. Within minutes they touched down at a vast open area. In front of them were several fighter ships. Small groups of aliens were advancing toward the fighters.

Daj set Luci on the ground near the perimeter of the small fleet. He worked with the data base to dis-arm an alarm system that surrounded the area then disabled the alarm on a specific vessel. They approached the ship and Minj sent a signal containing an entry code. A small, round door eased open. The opening was tight for Erik, who barely squeezed through.

Once inside they spotted a cockpit and a couch area. Erik was glad there were no individual seats given the size of humans compared with the aliens. Minj quickly navigated the data base to learn how to operate safety harnesses. The captain eased Luci into a lying position and secured her with a harness.

Daj looked at the instrument panel. It seemed to consist of various gages with readout screens only. There was no obvious way to engage or control the vessel. He began to wave his hands to interact with the shipboard systems. Nothing happened. He did some further reading from the alien database. "Fascinating. I control this ship with my mind." He was silent while a look of concentration crossed his face.

The fighter slowly lifted off the ground. As it picked up speed a round port in the great cavern's ceiling slid open. The ship shot out of the cave and into the sky. Several explosions lit the area and occasional fighter squadrons headed up from other underground silos. Erik was praying as they headed toward the blackness of space and stars began to appear. As the curvature of the planet came into view, a couple streaks of light shot past their bow.

"Warning shots?" asked the captain.

The other man nodded . "I'm sure they think we're ignoring orders, breaking formation . . ."

"Just get us out of here." The ship picked up speed.

Erik took over than hand-held. "Can't locate our ship, which is good. Means they're still cloaked." He ran some calculations to determine *Freedom's Hammer's* likely trajectory given the need to avoid the incoming navies as much as possible. Meanwhile, the beginnings of the alien war put on quite a light show. "Does this thing have a shield?" he asked.

"They don't seem to have that technology."

As they sped farther away from the planet the concentration of ships and firepower decreased.

Houston's stomach eased its grip. "Good thing a star skimmer accelerates gradually. Otherwise, we'd never have a shot at catching our ship," he said.

He tried to contact the ship on his comm link but kept getting static. "Dang. We need to let them know we're coming."

Minj nodded.

The captain continued working with the comm link for some time. Nothing. "We need to figure out the alien's communications system," he said. "Our people will think we're the aliens."

"And shoot us down?"

"You got it."

30

The Bridge

ONTOYA WAS FINALLY BEGINNING to feel safe as *Freedom's Hammer* continued to pick up speed. The humans had gotten past the two competing navies and even an occasional straggler ship.

"Sir, there's an alien vessel pursuing us and it's closing in fast," said a crew member.

"Get me a visual." snapped Montoya. As he viewed a 3-D of their pursuer he stroked his beard. "Why would they bother to send one measly fighter after us?" He paused for a moment before giving his next order.

"Communications, send them a message. Ask them what they want."

"Aye, sir."

Minutes passed but there was no response. The vessel was getting closer. "I don't like this," said Fred. "Fire a couple warning shots then if that doesn't work we'll blow them out of space."

"Yes, sir."

Several moments passed before a crew man announced: "They've slowed down a little but are staying on course."

The new captain slammed his fist into the wide chair arm. "How did they ever find us? Is our cloak still functioning?"

"Aye, sir."

"No matter. Prepare to fire laser cannons on my command."

Abruptly, a large, lighted area appeared in place of the port. It was a fuzzy 3-D image of two men. "Attention, *Freedom's Hammer.* Hold your fire! Repeat, hold your fire! It's us, Captain Houston and Daj."

The bridge erupted into cheering.

Montoya smiled from ear to ear. "Roger, Captain. It's great to see you. Is Luci with you?"

"Yes. She's still hanging in there." The view panned to show Luci, unconscious and lying down.

"We would like to rendezvous and should be in docking distance within ten minutes," Erik continued.

"Excellent, Captain. Welcome home. Crew, de-cloak us and prepare the landing bay."

31

Reunited

CAPTAIN HOUSTON AUTHORIZED TRANSMISSION of a broadcast to all inhabited star systems announcing that their crew had been to the aliens' planet and had discovered vital information that would allow humanity to better ascertain and respond to the potential threat. The message was general and used the wording headquarters had mandated upon successful completion of the mission. That would enable the nearer star systems to take relief and comfort from the humans' in-tel victory years sooner than if they had to wait for official communications from the regional commanders. Just as word of the original terror and the second attack had gradually spread among widely scattered worlds, the badly-needed word of hope would circulate just as quickly. After that transmission the ship beamed a coded, more detailed message back to base.

As the vessel put millions of miles between it and the interstellar battlefield, Houston lay on his hover field, exhausted after the two most intense days of his life. All the recent events kept playing through his mind . . . the traps the aliens had set at various points in space, the trauma of finding the nearly destroyed A'laaman ship, the shock . . . and pleasure . . . of seeing Zama again, the lack of sleep, the tough battle on the hostile world. He was amazed to still be alive. He hurt over the loss of the fifty-four A'laamans who didn't survive the trashing of their ship and the six of his own crew members who had died during the planet-side battle. But the scene that kept playing most vividly in his mind was Zama getting wounded.

He had comm linked Montoya so many times the last several hours for updates on her condition that he had angered the physician, who had also not rested the past couple days. The captain finally gave up trying to unwind and rolled to a standing position.

He walked down a quiet hallway to the infirmary but found her snoring lightly. He sat in a chair and waited. Her face was drawn. He continued to sit by her side. Montoya came and checked the beeping electronic monitors. A med tech administered an IV. An occasional A'laaman or two entered and stayed for several minutes out of respect for the former chief exec and appreciation of her bravery. The patient continued sleeping. Finally, Erik got up and took a long walk down the hall. He circled back to Zama's room. She was still asleep. They were alone in the room. He stood by her side and held her hand. The slumbering patient showed no reaction.

He let go of her hand. He went around the corner and checked on Luci in Recovery Bay 2. He left the infirmary and headed to Daj and Lisa's quarters. Along the way, several A'laamans stopped to congratulate him on the mission's success and to thank him for protecting their lady exec and their other makeshift troops. Erik commiserated with them on their losses. When he reached the Minjs' quarters, he spent some time talking with the couple.

When he left, the door closed behind him. He thought of Daj and Lisa's love for one another. The captain felt as if he had been dropped down a lift shaft. He suddenly felt like he was the only person in the universe.

After tending to some duties and meeting with his senior staff, Erik got back to sick bay and looked at Zama. Her eyes were mere slits. He softly spoke her name. She opened her eyes wider. Those orbs were a little glazed. She turned her head toward him. Suddenly her eyes flashed with recognition.

She held out her arms for a hug. Aware she was in pain, he approached her gingerly.

"C'mere, big guy," she said, grabbing onto him and pulling him close. She held him tight. "Thank God!" she said. "I thought I'd never see you again."

"What about your shoulders? I don't want to hurt you."

"Stay right where you are, mister." She burst into tears. They held one another for a long time.

She seemed too overwhelmed to talk more and he wanted to give her a chance to rest so he left the room.

He came back the next day. "How're you feeling?" he asked.

"The meds have taken the edge off."

"Anything I get you?"

"Cup of water."

He poured some water from a self-cooling pitcher that was covered with condensation. She eased up on one elbow and gulped down the beverage.

He continued: "I feel so bad for you. Your people put a lot of effort into building that ship . . ."

She sighed. "Minj and I put years of our lives into it. So did a number of other people. It's like losing a child. But . . . losing it gave me a chance to see *you* again."

He smiled. "But what about the future? I *don't have a way* to get you back home. This ship belongs to our Association and they may have another urgent mission for me."

She nodded. "I suspected as much. So . . . no way back." Her voice broke.

He looked into those brown eyes. "I'm sorry. If these were normal times I'd be able to plead your case. That would likely move all of you up in line for a Second Contact and you could hitch a ride on that ship but it would still take several years until launch, best case scenario. But ever since the aliens wiped out two of our colonies, everything has changed. It's all about increased defense spending and preparation for potential war so the contact missions have been put on hold.

She swallowed hard. "Years have already passed on A'laama since we've been away. Life sure throws you curve balls. I was once leader of a planet. Now I don't even *have* a planet to call home."

He stood up and clasped her hand. "You and your people will always have a home on our headquarters planet. Once we get back, I'll see to it that you're all well cared for."

She rubbed her thumb over his knuckles. "You're a good man, Erik Houston." she said.

The captain felt a hot rush in his face as he headed out the door.

32

Survivor

LUCI HAD BEEN SITTING on a bunk in sick bay, staring at the wall. She had ignored Montoya's orders to get some sleep. Even the meds hadn't gotten her to unwind. When her best friend entered the room, she did not react. Marji stood at in the doorway for some time then finally took a seat. The patient continued to stare at the wall.

Finally, she said in a low voice: "Not all of them are evil."

Marji cocked her head.

The patient turned toward her visitor. She gasped at the condition of Luci's face.

"Not all the aliens are evil," she repeated. "One of them interrogated me. He pushed his ugly bug face to within inches of mine. He could read my thoughts! And I felt like he could see right through me. I've never been so petrified . . ." Tears started to flow. Marji stepped over and held her friend.

Several minutes passed before Luci was able to continue. "But," she finally said, almost a whisper, "while he was probing my mind, I got a peek inside his, too. I can get a mind scan when I get back to Headquarters and maybe they'll find something useful about the aliens."

Her friend nodded. "You've been through a lot. I would have been terrified. But . . . you say not all of them are bad?"

"That's right. One of them saved me from a snake before it could strike me. And later, I think it was the same alien who tried to hold back the ones who wanted to hurt me. I think he was trying to help me, but they forced him to leave. Surely they have a spiritual life, the chance for a relationship with God like we do. There must be . . . *some* other good ones."

Marji pursed her lips. "Maybe so," she said. "It would make sense. The answer might be on that disk we're taking back to headquarters."

The women said nothing further for several minutes. Then the wounded lady spoke again. "I've asked two or three times for a mirror and they've never given me one."

The other woman's brow furrowed. "I don't think . . ."

Luci snapped: "I can take it. I want to see what I look like."

Marji sighed. She put her hands together then pulled them apart several inches then made the same motion perpendicular to the first. A square mirror appeared in Luci's hands. She glanced at the glass and flinched. Then she took a long moment to study her reflection. A wide pink scar marred the right side of her face, similar to the scar on her right arm. Her right eye was dark and swollen almost shut. It throbbed due to an infection but the doc had said she wouldn't lose the eye. "Well, my *left* side still looks good," she joked. "Should I wear a mask over half my face?"

Marji smiled a little. "Montoya's the best doc we've ever had," she said. "Once those flesh regenerating cells have finished working, you'll be turning guys' heads again."

Luci smiled, shooting an arrow of pain through her head. The smile faded. "Once we get planetside, you owe me a double date."

"A double . . . I *don't know* . . ."

"C'mon. It'll do you good. And it'll do me good."

"All right. You're on, sister."

Suddenly, Luci's face crinkled in agony.

"What's wrong?" cried Marji.

"They're still here," she replied through gritted teeth as tears began to run down her face.

"Who's here?"

"The aliens."

"But they're all dead. They're in cryo bags and locked in storage."

Luci shook her head. "You've got to get out of here. Get back to your post. Now."

"Will you be all ri . . ."

"Just go! Hurry!"

Marji hustled out of sick bay and back toward Engineering. She got there just ahead of the captain, Montoya and Irv who each burst into the room, their side arms drawn. Two aliens, one with a hand-held device stood in front of the department computer. The star men vaporized them.

"What does it take to kill them!" cried the captain. "We thought they were all dead. Fred, see what you can find out about their biology. I'd just as soon jettison their carcasses but headquarters wants to dissect them."

"Yes, captain."

"Faubner," Houston continued, "get your people to check out this system and make sure these creatures didn't do any damage."

"Affirmative."

"Irv, I don't want any more surprises. From now on I want a 'round the clock armed guard watching their frozen butts."

"Aye,sir."

Monotya conducted autopsies on several of the creatures. One of them still showed dormant signs of life. Hours later, he reported: "Captain, the aliens are a hardy species. When they're under extreme duress, their metabolisms nearly shut down and they go into state that mimics death. Some of the bodies in cold storage may still be alive. We'll have to examine each one."

The captain verged on giving the order to jettison. "All right," he sighed. "Luci's still in recovery. Get whatever personnel you need, do the exams, and report back to me."

"Yes, sir."

The physician continued his work. One of the supposed corpses moved an appendage a few inches. An armed crewman quickly dispatched the creature. Two others of the species still showed signs of life so the crew ensured their demise as well. The doc and his assistants returned each specimen to its bag and returned them to the cryo tank. Irv put the armed guard in place.

Marji and the other engineers checked and re-checked the ship's systems and everything seemed to be functioning normally. They found no evidence of viruses or sleeper problems.

The captain breathed a sign of relief as the threat from the aliens seemed to finally be over.

But three days after the humans left the planet, the Aliens' Revenge hit. All humans who had been directly exposed to them became violently ill with severe flu-like symptoms. Montoya was among the hardest hit so he was too ill to study the ailment clinically or work on an antidote. Erik and Zama were sick at the same time so they were unable to care for one another. As the first wave of humans was beginning to recover, the illness began to spread through the other inhabitants of the ship. Halfway through the second wave, the doc was able to develop a remedy. As the last patients began to recover, Erik wondered what else could possibly happen on the remainder of the trip back home.

33

Relating

AS THE SHIP CONTINUED its progress toward the regional base, it passed through a vast stretch between solar systems. The vessel came within 100,000 miles of a wandering planet. The dark world was an isolated, frozen rock that drifted through space, unaligned with any sun.

Aboard ship, Zama was soon able to leave sick bay and return to sharing a room with the recovering Luci. The A'laaman woman was a natural politician and had made friends with most of the crew.

Her best friend was a woman who had visited her when she was in recovery. Zama remembered her initial impressions. The crew member looked as tall as Erik and had long, red hair and freckles. Zama had seen her around the ship but had never previously met her.

"Hi," said the visitor. "I'm sorry about what happened to you."

"Thank you."

"I'm Nova."

"Zama."

She handed the patient a cloth bag.

"What's this?"

"A couple tops. Figured you could use them."

Zama nodded. In addition to wounding her, the alien attack had burned off her shirt above the chest. The remnant was only good for scraps. "That's really thoughtful of you."

The visitor shrugged. "They might not be a good fit but I'm probably as close to your size as anybody. Well, hope you're soon feeling better. And if you ever get tired of the food here, let me know. I have connections." She winked, turned on her heel and left.

Nova haled from a planet of even lower gravity than Zama's yet on her planet she was considered short. She worked in hydroponics and food production. So far, the A'laamans' hitching a ride on the ship had not resulted in a food shortage. The crew member supplied her friend with her one guilty pleasure, a small sweet roll that was her favorite star man dessert. Zama hardly conversed any more with Virtual Kelly.

She ran into Erik frequently on the crowded vessel. Now that the mission had been completed, the captain had more free time and began to spend much of it with her. Before long they were laughing about how they had worked together on her planet. Old feelings began to come back, gradually overcoming the relationship challenges caused by the time differential.

One day, the two of them happened to be in the gym at the same time. She was in a different outfit than normal, one she had borrowed from Nova. The clothes consisted of shorts and a sleeveless T-shirt. The shorts were a little tight and the shirt revealed part of Zama's midriff. She would occasionally tug at the clothes in an attempt to get them to cover more skin. She caught Erik watching her a couple times. She walked over and slapped him playfully on the shoulder. "Pay attention to your own workout, captain," she teased.

Whenever possible, the two of them ate together, using the star man utensils that consisted of a few small cylinders that each created energy and levitation beams that allowed one to size, cut and move a bite-sized portion of food from a plate through the air to the mouth. He had given her a set of the implements years ago when he visited her planet and she had used them on occasion. Since she'd been on his ship she had become quite adept at using the tools without dropping morsels of food on her clothing. A typical meal for her consisted of a small piece of fish or meat, a leafy bowl and a flagon of water. She also drank juice but rarely used mild stimulants like wakebrew. She also spent considerable time in the gym.

One day as the couple was in the lounge eating, the captain pulled away from his companion. "Lt. Faubner, are you *sure?*" he cried. He paused for a moment to listen to a response in his earpiece. He turned his attention back to his lunch partner "I'm sorry, Z. I've got to go." As he stood up, she waved goodbye.

Zama bit her lip. As the captain hurried off he wore a grave expression, one she had not seen since the battle with the aliens.

34

Stranded Light Months from Home

THE CAPTAIN RACED DOWN the hall. It would be a long time before he would have another spare moment, especially for personal matters. He double-timed it to Engineering. He burst through the doorway to see his lead engineer and the other techies all looking as if they were at a funeral.

"Tell me that again," he said.

She turned and faced him. "We're slowing down too quickly. 'Way too quickly.'"

He scratched his head. A ship that traveled at close to light speed took several weeks to accelerate to full velocity and an equally long time to slow back down prior to arriving at its destination. "Specifics, lieutenant."

"Here's our present speed," she said. She waved her hand and a six-inch high string of florescent red numerals appeared in mid-air. Several of the furthest numbers to the right were decreasing so fast they were almost a blur. She waved her other hand and a 3-D of the surrounding interstellar space appeared. The topographic map showed the warping of space caused by the gravitational pull of individual stars and planets. Houston recognized the red giant from the aliens' star system, several other nearby suns and their ship's destination, home base. A yellow dot showed the vessel's current position, more than ninety per cent of the way back home.

"The rate of slowdown is increasing," she continued. "I've extrapolated the end result and project that we would come to a complete stop here." She pointed her finger and a blue dot appeared. Houston whistled. She was picturing them being stranded too far outside a solar system and even too far from any interstellar station. There would be nothing around but the blackness of space for over two trillion miles.

"What's causing this?" he asked.

"We've checked and re-checked our propulsion system, all our other components. There's nothing wrong with any of them."

"Then what's going on?"

"We're slowing down because our system is telling the ship to do so. But we can't find anything wrong with it, either."

The captain lowered his head and rubbed his eyes with his thumb and forefinger. "It's those aliens again, isn't it? They're more trouble dead than they were alive."

She nodded. "That day we caught two of them with that hand-held device, we ran every check imaginable on the system and couldn't find any viruses. Everything still worked as it should. They must have uploaded a worm."

He looked her in the eye. "We don't have a lot of time to fix this. I want everyone working on it. Even Minj."

Her jaw tightened. She had always considered Daj a rival and had been jealous of him on their old ship, the *Initiative* when he'd been promoted to chief engineer. She had felt relieved when he'd stayed behind on A'laama. "Yes sir. I'll . . . get him involved."

"And lieutenant."

"Sir?"

"This mission has been tough on you and we couldn't have succeeded without you. When we get back to base I'll recommend you for a promotion."

"Thank you, captain." For just a moment, the furrows disappeared from her brow.

She kept the declining red numerals and the 3-D map in view to remind her crew of the urgency of their plight. Her research took her through a variety of data as she attempted to find any possible cause… and antidote to their dilemma. She even studied the records of the massacre of the Rantran colony years earlier and the next attack on a colony several years after that. She shook her head. The number of alien ships recorded in the second attack was 94.6 percent of the number in the Rantran incident but the number recorded by her own ship when the alien fleet had arrived at its home planet was only 70.8 percent of the number in the Rantran episode. Records from the second attack indicated the humans' ambush had destroyed about five percent of the alien ships. The humans had taken out an additional eight per cent of the fleet in subsequent attacks. Other than that, how did the size of their navy keep decreasing? It didn't seem likely the species would scale back the number of ships they would send to an all-out war defending their home turf. So how . . .

She thought back to Rantran. Records showed the lunar sub-colony had a nanobot cloud in place at the time the attackers had reached the lunar habitat. Her jaw dropped. Was it possible the colonists had programmed the bots to board some of the ships, plant a virus and eventually bring down a number of the vessels? She smiled. She had no hard evidence but perhaps the colonists had had the last word after all. She had analyzed the records from the next encounter with the species several years later, the broadcast Luci had recently picked up. The info did not include that level of detail, but did that group also use nanos against the invaders? Maybe the bots' subtle counter-attack had even reduced the number of alien vessels sufficiently enough to influence the outcome of the war with their mortal enemies. The humans' strategy had taken years to execute, perhaps due to discovery and counter measures by the insects, but the humans had ultimately succeeded. But wait! Her knees suddenly felt weak. The human-inspired virus had stopped, if not destroyed, a number of the alien ships, so . . .

The chief engineer set up a sound block and comm linked Erik on a private channel. "Captain," she said, almost a whisper, "our ship's speed is the least of our problems. We need to keep an eye on our life support, we need to..."

His fists clenched so intensely his knuckles turned white. "Yes, ma'am. Yes, ma'am. Stay on it, all of you! We've got to get this solved!"

But as her crew plus Minj continued their efforts they remained unable to isolate the worm or slow the deterioration of their ship's speed. They found nothing wrong with any of the ships' life support functions or any of the other systems. For now.

Over the next couple days, velocity continued to decline with no apparent solution. Marji strode toward Erik. "Sir," she said. Her hair was pulled back, but strands and pieces stuck out in all directions.

He turned around and looked at her. "How are we progressing?"

She sighed. "We're not. We've exhausted all possibilities." Her voice sounded flat.

"Check the data again."

"We've checked it and re-checked it. "

The commander frowned. "You're saying there's no solution."

She shook her head. "Unless . . ." Her eyes showed a spark of light. "We need to look into that alien database."

"It's worth a try. Go to it."

But even extensive searches of the alien info didn't yield a way to locate the worm or undo its damage. Among many random facts Marji learned is that the ant people had never known about cyber warfare until they learned it from the humans via the Rantran nanobots. She yanked on her hair, pulling out several strands.

Half a day later, the crew was finally able to isolate and disable the worm. The team then got the ship back into acceleration mode. The vessel continued to pick up speed and all seemed well over the next day. As a precaution, the chief engineer continued to closely monitor the speed. Late one night, a silent alarm awakened her. She rolled off the hover field in her night clothes and threw on a robe before rushing to Engineering. She frowned upon arrival at her post. She had set the alarm to go off only if the ship had again begun to decelerate. The still-present red numbers confirmed her worst fears. They were again getting smaller.

She sent comms to awaken her two assistants but allowed Minj to continue sleeping for now. The crew stayed up all night but their analysis was the same as before: the engine was fine and there nothing physical causing the deceleration, nor had there been any further tampering with their computer system, at least as far as could be determined. The chief engineer rapped her knuckles on a counter. She was determined to get to the bottom of this. Another day of fruitless work passed.

Too tense to sleep that night, she sat in a hover chair and propped her feet on a counter, relaxing as much as possible. Eventually, her eyes grew heavy and her mind drifted off. In the depths of the night, the lady snapped awake. Someone was moving around in the darkened Engineering section. She snapped her fingers and light flooded the room. Standing in front of the computer was a short figure who ignored the engineer's presence while typing furiously into a hand-held device. The invader's eyes were glazed.

Marji's stomach froze. Luci *was* the worm!

The officer sent an emergency code to the captain, Fred and Irv. Marji ran and tackled her smaller friend, knocking her away from the console. The women struggled and the hand held dropped on the floor. Luci pushed the engineer aside and grabbed the gizmo. The men dashed into the room.

"Stop her!" Marji cried. "The bugs have screwed up her mind."

It took all four people to hold the struggling woman. Her fingernails tore across Montoya's arm, drawing long scratches that filled with blood. Her teeth chomped into the captain's face and he almost dropped her. She kicked Irv in the stomach. Houston retrieved a weapon, set it on stun and fired. It took three shots before the attacker eased into a motionless, heavy mass.

35

Secrets Deep in the Mind

THE GROUP CARRIED LUCI'S limp body to the OR, laying her on her back. The doctor waved his hand and over-sized scans of various parts of her brain appeared on the walls of the room. "Get me a couple med techs," he growled.

He looked down at her and sighed. "What have they done to you?" he said.

His staff and he worked for some time but they were unable to isolate the source of Luci's trauma. Finally, he comm linked: "Captain, I can't figure this out. She may need a specialist planetside."

"Keep her sedated. I want an armed guard with her constantly."

"Yes, sir."

The doctor gazed down at his sleeping patient. She still looked tormented. He shuddered.

A few hours later while he stood by her hover field she snapped awake. She shivered yet beads of moisture gathered on her brow. She began to weep. He held her and tried to comfort her.

Sometime later when she was more sedate, Marji knocked on the door. The physician opened it. The armed crew member stepped aside to allow her to enter.

"How is she?" asked the visitor.

"Traumatized, confused . . ."

"Any luck with the brain analysis?"

He shook his head.

Marji gritted her teeth. "Somewhere deep inside her mind are the instructions the aliens had her feed into our system. We need that info if we're ever going to get home."

The doc sighed. "I don't want to just randomly poke around. She's been through enough."

She laid her hand on Fred's shoulder and looked him in the eye. "I love her like a sister. I don't want to hurt her either. But we've got to figure this out."

He turned his gaze back to his patient. "I hate to do this, but I hope you can take just a little more."

Her pale face held no expression.

"Would it help if I stayed here?" asked the chief engineer.

"Don't you have . . . things to do?"

She looked the floor. "I've hit a wall. And I think my head needs a rest."

"Your being here may help her."

The visitor grasped the patient's hand. "Luci, this is Marji." she said. "Can you hear me?"

There was no response from the patient and her vitals remained unchanged.

"Sweetie, we need you to relax and stay calm while Dr. Montoya looks for something the bug people have buried deep in your mind. It's really important that we find it. You've been brave and you've been strong. I know this is tough but Fred and I are going to help you through this, okay?"

Her friend's blue eyes continued their vacant stare. The engineer nodded to the doctor, who began a deeper brain probe.

Minutes later, Luci clenched the other woman's hand and let out an agonizing scream.

Meanwhile, the ship's deteriorating speed dropped to barely above that of ancient chemical rockets from the dawn of spaceflight. At this rate it would take the vessel several thousand years to get back home. Soon, the ship stopped moving altogether. While the captain ordered a comm tech to send out a distress signal, the aliens' program moved on to its next target: life support.

36

Teach Us to Number Our Days . . .

ZAMA WALKED BY A female crew member who shot her an annoyed look. The A'laaman sighed and continued down the hall. Morale on the stranded vessel was at a low ebb and after several weeks of the star people and the A'laamans living together, the overcrowding was causing nerves to fray. She was grateful the spacefarers had accepted all the shipwreck refugees from the very beginning. But now she was starting to feel like a burden. Not only did everyone feel helpless because the vessel had stopped moving, they also knew it was a race between how soon the techies could de-bug the system and when the ship became uninhabitable. There were no secrets on a small, overpopulated starship. She hated to see everyone in such a negative frame of mind.

She had organized a few prayer meetings with small groups of star men and A'laamans. That had seemed to help morale. God willing, she hoped to live many more years but she was also not afraid of stepping into eternity.

She thought about when she had faced death once before. Years earlier the assassination attempt on her native planet had left her in severe pain for several days. The physicians had been unable to revive her when the last spark of life had faded. She had seen a vision of her mother welcoming her to heaven but Zama had not entered paradise, it had not yet been her time. Her miraculous revival had served as a witness to many people.

Her thoughts returned to the present. Her hand gripped a railing. *"Lord, only you know if this is our time,"* she prayed. *"But if it's your will, please get us all home safely."*

The temperature had dropped a few degrees and could no longer be controlled. The oxygen level had declined slightly, creating altitude sickness for some. Montoya supplied meds to treat the symptoms for now. Once the situation reached a critical stage, the star people would have a temporary stopgap, but one that would only work for some of the vessel's inhabitants. Headquarters had assigned the captain and each crew member a warm suit and oxygen mask that could keep them alive a little longer in the event of life system failure. The ship carried a couple extra suits and the star men had lost several people in the battle on the alien world, but there were still not enough suits to cover both the star-men and the A'laamans. In addition, the slight height and proportion differences compared with the space people made some of the A'laamans difficult to fit. The situation further added to the tension between the two groups.

While Zama had all of this on her mind, she stared out a port at the black, starry sky and bit her lip. "Hey, Z-z. Why so down?" asked a soft voice. She turned around to face Nova. "Listen, I-I want you take my suit," the red head insisted. "I'll bet if will fit you or at least pretty close. And I'll give you the oxy, too."

The A'laaman looked at her and shook her head. Her friend appeared to be about twenty-five, even younger than Zama's kids. "I can't do that. I've lived a full life. A very good life. But you have your whole life ahead of you."

"Please. I want you to take it. Besides," she added, lowering her voice, "Captain Erik has his eye on you. It would just kill him if you . . . if you . . ."

The other lady blushed and lowered her eyes. "The captain and I have . . . a history. We enjoy one another's company, but that story has yet to be written."

"Well, either way he cares about you and I know he'd want you to live."

Zama turned back to looking at the stars.

"I hope you don't think he put me up to this," said the crew member. "This is totally my idea."

"I'd rather go without a suit than know I was condemning you to death. Besides, there's a chance your people can solve this before it gets that far, right?"

"I hope so."

Similar conversations had been taking place in other parts of the vessel.

37

Too Stressed for Sleep

THE CAPTAIN WAS HAVING another sleepless night. The lights were out on the bridge and he was pacing the floor. He walked over to the port and looked out into the blackness of space. He felt drained and helpless. If Faubner and her crew couldn't soon find a solution it would all be over. He had had meetings with Irv and the other officers. They had discussed all possible ways to keep everyone alive as long as possible. The ship had sent the distress signal several days ago but the nearest starbase was weeks away. This had been the toughest assignment of his life. Most of his people had survived and now it was all going to end like this? His love, his Zama had come back into his life and now they could both perish together? Life, the cosmos, didn't seem fair.

"God, why are you doing this to us?" he whispered.

Hours later it was morning, shipboard time. He saw her walking down the hall and motioned her over to him. He set up a sound block. She had dark circles under her eyes. He placed his hands on her shoulders. "How are you?" he asked.

"I've been through tough times before."

"How are you feeling?"

She looked at the floor. "I've felt better."

He cleared his throat. "Nova told me . . ."

She nodded.

"I, uh . . ." he stammered.

"I know you didn't ask her. She's a sweet person."

"One of our best. When we get back . . . when we get back, I'm going to recommend her for officer school. And . . . I want you to share *my* oxygen. Nova can have her own."

She shook her head. "I-I can't do that. You're the captain. You need to survive."

"All lives are important. I'm responsible for every life on this ship. And *your life* means a lot to me."

She swallowed hard.

Their eyes met. Hers were watering. His arms encircled her. He held her and stroked her soft hair. "It will be all right," he said in her ear. "We'll get through this. We'll all get through this."

She pulled him closer. "Stay with me, captain," she whispered. Her warmth against him felt comforting. They held one another for a long moment then eased apart.

After mess call, he held a meeting in the lounge for all personnel, both staff and refugees.

"This is the decision I've come to," he said. "For every one of my crew members whose suit fits an A'laaman. I've asked that they give that suit to the A'laaman. Any woman who does not have a suit and can fit into a suit designed for a man should take that suit. I deeply regret that there are not enough suits for all, but those are my orders."

An enlisted man protested. "Sir, every one of these people are our friends. We can't sit around and watch them die. If there aren't enough suits for all of us, then no one should have them."

The room erupted into loud talking and shouts. The captain raised his hands for silence. The din continued.

"Let's have order, please. Order!" shouted Irv.

After a few seconds the remnants of conversation died out and the captain was able to continue. "I appreciate what you're saying. This tears all of us to pieces. But some of us surviving is . . . is better than no one. We've had too difficult a fight to give it all up now. Lt. Faubner, Daj Minj and the other engineers will keep working to find a solution to our dilemma. But my orders stand."

The room lapsed back into animated talking. Under the captain's rules, three A'laamans and two crew members would be without suits.

38

Ghost Ship

"SIR, WE'RE PICKED UP a distress signal from a disabled ship!" the comm tech shouted to the captain.

"What are the coordinates?" snapped the commander.

The tech responded with three sets of numbers, one for each dimension. "That's a week away from here. Get me a visual."

"Yes, sir."

A 3-D of the vessel appeared in front of the captain and slowly rotated. The ship was totally dark.

"Looks like they've lost all power. There's not much chance of anyone surviving by the time we get there."

"But we can't take that chance," said his second in command.

The leader was silent for a moment then slowly nodded. "Flight crew, plot a course for the following coordinates and advance at full speed . . ."

The pod arrived at the motionless ship but was unable to get the airlock open. The rescue vehicle used plasma torches to cut its way in. Once the transport was inside, several space suited humans hurried into action. They were able to get inside the inner airlock door. They used light wands to find their way down darkened halls. They had barely begun scrambling when some faintly-lit, red 3-D letters about six inches high greeted them with a message: *All personnel are on bridge. Medical emergency. Please hurry.* An arrow pointed the way.

The visitors ran down the dark halls as fast as safety would allow. As they burst through the open bridge entrance, they saw another sign, its letters beginning to fade: *All personnel are in medically induced, death-like state. Do not assume death. Make all possible efforts to revive.*

They noticed about twenty bodies, most but not all clad in warm suits. Some were slumped over, some lying on the floor. Their exposed facial skin was cold to the touch but their core temperatures had not yet dropped much. The leader saw the shoulder patches and insignia on one man's warmsuit. "This is the captain," he comm linked to his crew. "Work on him first."

Then he spotted the woman slumped over near the commander. The lady had an oxy mask hanging from a strap around her neck. Her face was blue. He motioned for the other med techs to hurry over to her.

He shook his head at the tragedy. If they had gotten here much later it would have been for naught. About a week earlier, the passing merchant ship had picked up the emergency signal then changed course to rush to the ailing vessel.

It was not a formal law but on most civilized planets, people were taught from childhood the Good Samaritan principle that stated: "Anytime you're in space and you pick up a distress call, you are obligated to respond to it. Because of the vastness and isolation of space, and the limited number of Space Patrol vessels, the passing ship may be the afflicted party's only chance of survival."

The vast majority of space travelers adhered to the teaching, most captains and crew realizing someday *they* could be the ones needing emergency aid. So, although occasional pirate ships lurked in some sectors and they had been known to fake emergencies, standard procedure was to evaluate the legitimacy of the call then proceed with caution but by all means don't risk failing to render aid.

Zama felt unable to wake up. She couldn't get warm and her head felt like it was splitting. Two days had passed since they had been rescued. She had spent the first day lying in a hover field with several IV's attached. She had spent the next day mostly sitting, wrapped in a blanket and staring at the wall. At times she would fall asleep, awakening later suffering from cramps from slumping over in an awkward position. During her waking moments, her aching body felt like lead and her mind seemed full of cosmic dust. She compared herself to a 120 or 130-year old woman like her great-great grandmother had been while Zama was still on her native planet. Her eyes saw Erik several times but at first she didn't recognize him.

As she sat on a hover chair holding her aching forehead in one hand she heard some muted buzzing. She concentrated. Two deep voices were talking in hushed tones. The volume was low enough to be irritating but too low for her to make out the words. She turned her head and squinted. She made out the forms of the captain and Montoya. When they saw her look at them they stopped talking and lowered their eyes.

"So . . . what's going on?" she asked.

Silence.

She bit her lip. "C'mon guys," she said. "We've all been through so much. There shouldn't be any secrets."

The two men looked at one another. Finally, Erik spoke. "Everyone is still alive," he said.

"Still alive. But . . . how is everybody?"

Montoya chimed in. "Three of your people including Captain Mies and two of our people are in comas. I knew I should have diluted some of those dosages."

"Easy, Fred. You did your best," said the captain.

Zama lowered her head. "And Nova?" she said, her voice breaking.

Neither man answered for a long moment.

"She went without oxygen too long . . ." said Fred

Her eyes bore into his. "Give me the full story."

"She's brain dead. But I've conferred with the doc on this ship and if she can hang in there a little while longer and we can get her to a specialist . . ."

Zama lowered her head. Her hair fell in front of her face. Her shoulders jerked up and down while her mouth made little sobbing motions. Houston touched her hand.

She did not look at him. "Honey, I . . . really need to be alone," she sniffed. She eased to her feet and shuffled down the hall.

"Will you be all right?" he called after her.

"I don't know," she said.

The news had cut through Zama's heart. She had no appetite for the next few days. The depression made it all the more difficult to get her body and mind out of their sluggish state.

A few days later she was feeling closer to normal and stood next to Erik staring out the port. A magnetic field from the freighter was pulling along his ship, which looked like a toy in the distance.

Montoya sat nearby, his head lowered and his hand on his head. "I could have done more. If I would only have noticed her mask had fallen off," he mumbled.

Marji had her hand on his shoulder. "Doc, you did everything you could," she said in a low voice. "Look how many lives you saved."

She continued standing beside him for a few minutes then walked over to the couple. No one spoke. Marji's hair was in a thin, tight braid. She stuck the end of the braid in her mouth and began chewing on it.

"Those devils aren't done with us," she said in a low voice.

"Who's not done with us?" asked Zama.

"The aliens."

"Not done with us? They almost killed us."

"But they haven't succeed. Yet."

"What do you mean?" asked Erik.

"Think about it. These creatures are thorough. We had to pass through several different danger zones just to get to their planet. And back out of their solar system. Do you really think we've found all their traps?"

"They made us sick, turned Luci into a saboteur, stopped our ship, totaled our bio system," said Zama. "What's left?"

Marji's eyes widened. "What if there's a sleeper bomb aboard our ship?"

"What if she's right?" said Irv. "If the bug people blow up our ship, there goes our info disc, their fighter ship and their precious little carcasses."

"And we need them for med research," said Montoya.

"Wait a minute," said Erik, holding his up his hand. "I don't want to act based on hunches. But we can't take a chance. If our ship is destroyed that not only negates our mission but any explosion could damage this freighter as well. Looks like we need to send a team back to the ship."

"I'll go," said Zama.

"No. I want you to stay safe."

"I need something to do. I-I'm worried about Nova. It's eating me up just sitting around here."

The captain looked as stern as ever.

"Look," she continued, "we're all in this together. You'll need someone to carry all those alien specimens."

He sighed. "Okay."

When the captain held an officers' briefing on the subject, his key people all volunteered. "Thanks for your support," said Erik. He swallowed hard. "Most of the officers plus Zama and Minj will go. He put in a comm link call to the commander of the freighter. "Captain," he said, "I have a big favor to ask . . ."

Zama, clad in a warmsuit with a light torch on its helmet, followed Marji as they floated through the corridors of the darkened ship. The women were weightless, the vessel having lost its artificial gravity along with its power. A small propulsion system on the backs of their suits allowed each woman to maneuver through the halls, although the chief engineer periodically needed to nudge her inexperienced counterpart to keep her on course.

"Here's the cryo storage area," Marji said through the comm link, pointing to a doorway ahead.

She turned a knob and pushed the door open. The lack of power had long since caused the cryo equipment to quit working but the coldness of space had crept throughout the vessel and now served as a natural freezer for the alien remains. Each of the humans grabbed several of the bio bags then sailed out of the room.

"This is the captain," said Erik's voice in their ears. "'We've located a bomb and it's on a countdown to blow in less than five minutes. Repeat, less than five minutes. We need to vacate now! Daj, you got that data chip?"

"Got it. But captain, maybe the aliens' info can help me disarm the explosive."

"Do we have enough time?"

"We're cutting it close."

"Get on it. But we'll prepare to e-vac everyone else just in case."

Next, Erik called the captain of the freighter. "This is Captain Houston. We've detected a bomb aboard our vessel. Disengage your tractor beam. Repeat, disengage."

"Aye, captain."

The crew propelled through one hallway and down another as quickly as they could. Zama lost control and bumped her hip into a doorway, causing her to squeal in pain. She let go of two of the bio bags, which floated off in a different direction.

"You okay?" asked Marji.

"Yeah," she said.

The women retrieved the bags.

"T-minus three and a half minutes," said Montoya.

"Daj, how're you coming?" called the captain.

"Slower than I expected."

"At T-minus two minutes you haul butt out of there. We don't want to lose you."

"Roger."

The group continued to navigate the hallways. They approached the airlock door. Erik flashed his hand-held device. The door did not move. "It worked on the way in. Any luck, Faubner?"

Her hand-held did not work, either.

"T minus two minutes, ten," said Fred.

"Minj?"

"Don't think I'll get it, Erik. Got the database. Catch up with you shortly."

"Hurry, buddy!"

The captain and Fred each drew ener blasters and blew open the door. The group hurried through and piled into the pod.

"Wait for Daj," said the captain. "Where are you, pal?"

"Just about to the air lock." Ten seconds later he dove into the last seat on the pod.

"Ms Faubner, get us outta here," snapped the captain.

She did not take time to reply. The bay doors opened and the pod tore off into space. The tiny boat put precious distance between it and the ship while the countdown clicked away. The merchant vessel began to increase in size.

"Detonation in five, four, three, brace yourselves!" warned Montoya.

A blast wave jostled the pod, slamming Marji's head into the starboard port. The motion also jerked Zama in her seat, hurting her ribs. Fred was closest to Marji and grabbed the controls. After a few moments of struggling he was able to right the pod. A short time, later the boat slid into the open bay doors of the cargo ship. As the transport came to a stop, Erik turned off his force field harness and helped Fred treat the ladies' injuries.

39

Remote Outpost

ZAMA WAS THERE WHEN the end came. It took place in a hospital within a starbase, the most remote crossroads of civilization in the area. The isolated station had grown over the centuries from a scientific research center into a trading post and docking facility that could handle up to a dozen ships at a time. It sat outside any solar system and was its own political entity, although the regional government over this sector paid to operate a military facility at the site. Several generations of civilians had been born, grown up and died without having ever left the interstellar oasis.

The cargo ship carrying the star men and the A'laamans had stopped at the base, the freighter's destination. No commercial ship was heading back to regional headquarters for two more weeks but a military vessel was departing in a week and a half and its commander had offered to get Erik and the others back home.

Meanwhile, Zama got used to her new surroundings. She'd had several temporary homes over the past few months: the A'laaman ship, Erik's, the freighter and now this base. She was getting tired of being a space orphan but was grateful just to have a place to sleep, a shower, food and functioning life support. The day she arrived, she walked along part of the base's perimeter and found a transparent wall providing a window to the stars. She quickly spied the Ava star, the larger of A'laama's two suns. It looked so close she felt she could touch it. She put a hand up to the glass. It felt cold, like the coldness of space. A shiver ran up her spine.

Ava was so close, yet inaccessible. It might as well be on the other side of the galaxy.

She would normally be inclined to take long walks on the station, which offered more such opportunities than aboard a ship. But her ribs were still sore from the bruising they had taken when the blast had jostled the space pod and her hip was still a little tender from her injury aboard Erik's ship.

One day, she was sitting on a hover field in the micro apartment the locals had assigned her. It barely had enough space for sleeping, a closet and a restroom. A lighted panel with several buttons hovered in mid-air. It allowed her to call up 3-D news and entertainment from the base's web casts as well as programs broadcast from neighboring solar systems. Of course, all shows but the local ones were several years old by the time the signal reached the remote starbase. She zipped through the offerings but some didn't interest her and others were in foreign languages, although she had the option of a simultaneous translation.

She switched to an info channel about the base, a small city-state with several thousand inhabitants. The channel included some local news, mostly small town gossip. She perked up upon seeing a video of a mysterious flash in deep space a couple trillion miles away. The reporter speculated on the nature of the light show but indicated very little knowledge of the details. Zama knew the answer. She had been there.

Next came a 3-D ad for the base's handful of retail stores. They looked like mere holes in the wall. But shopping was another luxury she had missed during the time she'd spent among the stars. She had no star man money but she felt browsing would do her some good.

An armed Navy man who was even taller than she escorted her everywhere she went. The local brass had insisted on that because Erik's mission was classified. She'd learned the military part of the base had tripled in size the past few years and there were rumors the place was heavily armed in case of an attack by any non-human forces. Even merchant ships like the one that had brought her to the base carried some ener weapons these days.

She stepped outside the door to see her personal sentry on duty as usual. "Uh..sir, I need to do some shopping," she said.

"Yes, Ma'am."

She strode off in the direction of the shops and her escort readily kept up with her. As they drew close to the retail area, she noticed a man and woman step out of the way as she walked by. Another man walked up to her and handed her a clear, coin-sized hexagon. It flashed with tiny red and blue lights. The man smiled at her and walked away. One woman hurried across the street in an apparent move to get away from her. A man walked toward her on her side of the sidewalk and wouldn't yield.

He stopped walking just a couple feet before he would have collided with her. He glared at her until she stepped out of the way.

"Sir, that's not how you treat a lady," the Navy man called after him.

"She's not a lady," he called back.

The sailor hustled to catch up with the offender. He motioned the man aside and spoke with him a moment. The man turned back and approached Zama.

"I'm sorry, Ma'am," he said, looking at the ground. "It won't happen again." He hurried away.

"Thank you," she said to her defender. He nodded.

Puzzled by the behavior of these space dwellers, she approached one of the boutiques. A beefy man showed up in the entrance, blocking it, his arms folded. His eyes shot lasers through her. She stopped in her path.

She looked at the reflection in a shop window. The woman staring back at her was dressed in rags and she had dark circles under her eyes. Zama wanted to crawl under the walkway. These people all thought she was a vagrant. But she had to admit she looked the part. Her hair was longer than she liked and her locks were frizzed. The small amount of makeup she'd owned had blown up along with Erik's ship. Her face was drawn and she looked like she had aged several years, perhaps her body's accumulated reaction to everything she had lived through since the A'laaman ship had been destroyed.

She was wearing one of Nova's ill-fitting shirts and the same pants she'd had on the past three months, the same pair she'd had on when Erik had first laid eyes on her aboard his ship. The only pair she owned. The pants were now faded, stained and threadbare although she had sonic cleaned them daily.

One pants leg had a small hole in the knee. Her face reddening, she lowered her head. She was about to turn around and trudge back to her cubbyhole.

Then, from the corner of her eye she noticed a woman in a shop two doors down was smiling at her. She turned away, thinking the lady wanted to ridicule her. A moment later she glanced back. Now the shopkeeper was waving. Zama turned back around and shuffled away, the military man trailing behind her.

The other woman called after her and jogged over to her.

The refugee turned around. "What?" she said through clenched teeth.

"G'day, ma'am," the lady said in an accent that was new to Zama. "We have some nice clothing and other items you may want to look at. You're certainly welcome."

The outsider, looking at the ground, gradually raised her head. The clerk's eyes looked sincere.

The rag-clad refugee smiled. She followed the woman back to the shop. She sorted through several outfits. It looked like some of the clothes would fit her. Her benefactor finally spoke again.

"You're one of those people who were rescued from the stranded ship, aren't you?"

Zama looked down. The military had given her strict orders to say nothing about their experiences. The man with the weapon was staring at her.

"That's okay, honey," said the kindly soul. "I won't pry. There's been a lot going on to combat the alien menace."

The refugee bit her tongue. This store clerk would be surprised just how much first hand knowledge she had of the "alien menace." She stepped away and began to look at other items. She spent some time trying on outfits. She decided on two pairs of pants and two tops then approached the clerk. She dug into a pocket for the hexagon, hoping it would be enough of whatever interstellar currency it represented.

"Tell you what," the retailer continued. "You've been through a lot. You can have them."

The other woman's jaw dropped. 'Uh . . . Ma'am . . ."

The sales lady said: "You heard me right. Both outfits. On me."

"T-thank you," said Zama, feeling like she had suddenly gone from poverty to wealth. "This means so much to me."

"Do you need underwear, too?"

"Well . . . yes."

"Pick out a couple sets. No charge."

She hugged the clerk.

As she prepared to check out, the military guy waved a wand over each item, apparently to check whether they were bugged or armed or in some other way hazardous. He gave his blessing to the transaction. Moments later the A'laaman was heading out the door carrying a bag filled with her purchases. A short walk later, the sailor and she arrived back at the entrance to the over-sized closet that was serving as her temporary home. He stopped at the door while she stepped inside.

She was so thrilled she felt like shouting. Now she had brand new clothes that fit. She had never felt more grateful. She changed clothes, throwing the old outer clothes and underwear into a disintegration chamber and pressing the activation button. She felt like a new woman.

Her revelry was short-lived as she soon faced a much more somber situation. The base hospital was caring for those with med needs: Luci and those unable to function on their own ever since the crash of the bio system on Erik's ship. Nova, severely brain damaged, was still in a coma. A number of people took turns staying with her, holding her hand and talking with her. Sometimes it was the captain, at other times it was Montoya, Irv or various crew members. Zama came a couple times a day, often staying a few hours each time. Sometimes she would lay her hand on her friend's face or forehead. "It should have been me," she told the patient on more than one occasion, "It should have been me."

The lady's condition deteriorated further. The docs had her attached to several different machines, the most sophisticated on the base. One day, while Zama stood nearby, the crew member's vitals crashed, sending off high-pitched, beeping alarms. Several docs and med specs rushed into the room, ordering the visitor out. Time dragged as she awaited word. What felt like days later, one of the attending docs found her. She could read the outcome on his face. The pain in her chest hit her so hard she almost blacked out. The physician held her arm to steady her then helped her to a chair.

"Could I see her?" she asked quietly. The med professional allowed her to rest a few minutes then helped her back up. Zama, still wobbly, held onto his arm as he walked her into the room. She looked down at her friend, whose face looked serene. She stroked the woman's hair. "Put in a good word for me in the next life, huh?" she said. A confidant, a sister, a daughter was gone.

"When I've got solid planet under my feet, I'll make sure you have a beautiful grave marker. We won't forget you," Zama promised.

She alerted the captain and asked him to pass the word. Within minutes he burst through the door with Montoya following close behind. Fred brushed past Zama and looked at the instrument readouts. There was no brain activity, no pulse, no blood pressure, no breathing, no signs of life. The patient was already turning pale. "She's gone," he concurred. He hung his head.

Many people soon gathered for an impromptu memorial. Erik and various others tried to comfort Zama but she felt numb. After a while, she excused herself and launched into a full speed walk, followed as usual by her shadow, her official guardian. She turned on her heel. "Sir . . . for once can't you just leave me alone? Please."

He did not answer for a moment. "Yes, Ma'am. And I'm sorry about your friend."

"Thank you."

She kept walking for some time, finally ending up on that walkway that followed the perimeter of the station. She passed the wall-to-wall port that looked out on the blackness of space and the myriad of stars. At one point, she rounded a gradual corner where a huge port looked out over the space dock and the several different ships that sat in enormous service bays hundreds of feet below. But none of these sights were registering in her mind. "Oh, God, why did this happen? God, what are you doing?" she asked, almost blinded by her tears. She absent-mindedly tasted their salt.

Her feet sore and her legs aching, she eventually made it back to her tiny dwelling. The sailor was standing guard near her door.

"Glad to see you, Ma'am," he said. "We were beginning to worry."

She nodded. She stuck her palm on her doorplate. The door opened. She closed it and it auto-locked. She flopped down on her hover field. She felt alone, so completely alone.

40

Reflecting on the Impossible

Montoya was lying on the hover field in the military transport, glad this challenging mission was finally complete. He shuddered to think how close he had come to the end. He could have died on that hostile world. Or on his own ship when the life support had failed. He thought back to when his lungs had hurt and when he'd had to struggle for each breath. Right before he'd blacked out, he'd thought: "This is it." Instead he'd awakened on a different ship, cold, feeling sluggish, but alive. Why had they not all died in space, trillions of miles from nowhere?

"Do you think that was a coincidence?" said a voice in his mind.

He slowly shook his head. No, not a coincidence. And not Einstein's Law. There was something much bigger here. His thoughts returned to the planetside battle. "We should have died," he thought. Either the aliens could have knocked them off underground. Or the humans could have been massacred in the protracted firefight. Or the invading ships could have atomized them. "We should have died."

"But you didn't," said the voice.

A knife cut through him when he thought of Nova. He had finally stopped blaming himself for her death. But why did that sweet, young crew member have to perish while an old curmudgeon like him had survived? He shivered.

"What is meant to be, is meant to be," the voice responded.

"What do you mean?" he asked aloud. But only an awkward silence followed his words.

The doc rolled over on his side. He briefly fell asleep and had a dream as vivid as life. He was back on the planet, battling the natives, surrounded by fire. He snapped awake. His mind drifted back to other missions, other planets. He counted the number of times he had come close to losing his life over the years. More times than he'd realized. But he hadn't thought much of each incident at the time.

And how had Erik and Zama ended up in the same place after light years and, for her, planetary years of separation? Finding your long-lost love from another star system in a shipwreck? It was the most extremely improbable event he'd ever encountered in his dozens of missions to the stars. And what about Daj becoming a father of some of the A'laamans but not finding out about it until three centuries later? And what about . . .

His mind was racing. He rolled to his feet and burst out of the room, hurrying down the hall. One of the sailors, an enlisted man, saw him and saluted. He absently returned the gesture. He kept speed walking until he found a port. He stared at the stars, the jewels of space. *Who had made all of this?*

"You do exist," he said.

A loud chime in his ear jolted him back to the reality of spacetime. "Montoya, here." He said.

A life-sized, 3-D image of Daj Minj appeared. "Hey, buddy. We survived another mission."

"That . . . we did," the doc replied, wondering if his friend appreciated the full impact of those words. Fred set up a sound and visual block.

Minj continued: "If you've got a few minutes, I want to share something with you. It's a discovery I made back on A'laama. Well, not just me. I had a lot of help from one of the natives, Tanna Bern."

Fred caught his breath. "Bern. Didn't she try to kill their chiefexec?"

"She was part of the conspiracy but she actually saved Zama's life in the end. Zama pardoned her, which got her out of prison. She even publicly forgave her. It was all over the media. Made a lot of people mad at the time but it was the right thing to do. Tanna helped me build the starship. Now we're business partners."

The doc was stroking his Van Dyke beard so hard he pulled out a few whiskers. *Had he fallen into a parallel universe?* What other surprises awaited him? He agreed to meet his friend in the ship's lounge.

"What's up?" asked the doc as he strode into the room.

The other man looked both ways then sound and visual blocked. "Thought you'd want to see this," he said.

A giant page of equations appeared, forming a wall in front of them. The longtime first officer had enough of an engineering background that he understood most of what Daj was presenting. Minj made a page turning motion with his finger and a second page appeared. A third page contained ship specs. After several minutes of silence, he snapped his fingers. The info disappeared.

"Do you realize what this could mean for the production of starships?" cried Montoya.

"Absolutely! We call this the Minj-Bern Principle. It's creative protected on A'laama. But we want to take it to a larger audience. Now that our ship has been destroyed and we have no way of getting back to A'laama, maybe it's time to start marketing our Principle. Then channel the eventual wealth back to A'laama so they could build a replacement ship."

Montoya nodded. "I've been doing a lot of thinking and could use a sabbatical. Maybe even a career change. Could you use a partner?"

"I couldn't think of someone better." The two men gripped hands.

"And this idea of your spreading your intellectual property among the stars knowing it will eventually bring a lot of visitors to A'laama, your . . . Madame Zama is okay with that?"

Daj was grinning. "It was her idea."

41

Zama on the Edge

ONCE THE STAR MEN and A'laamans returned to Erik's planet, the military quarantined Zama for three days of debriefing. One day, she'd had to endure a mind probe that had created little voices in her head and the naked humiliation of feeling strangers prowl around in her innermost thoughts. One official had even verbally asked about her intentions toward Erik, a question she found intrusive. After being released, she learned that all the other natives of her planet had gone through similar experiences, minus the questions about any potential romance. The interrogation and the mind invasion had left her drained. For the next few days she felt angry, disoriented and violated.

Meanwhile, the public began to clamor for access to their new heroes. Once the leaders began to allow news coverage, the returnees were media sensations for weeks. The star men made certain the e-press was aware of how the A'laamans had helped fight the aliens and had lost some of their own in battle. The City of Newhattan threw a parade in the honor of both groups. The media broadcast the festivities all over the solar system and beamed them to the stars.

Zama had had her share of media attention but the local political and military bosses had strict guidelines on what she could and couldn't say. She found this disconcerting. Years ago, when she had been the ruler of A'laama she'd felt reasonably free to say what she wanted. She had begun to feel isolated and discouraged when she had been on the space station. Those feelings continued now that she was planetside. She was struggling to adapt to the local culture. It was so different than her home world. Newhattan was five times the size of her capital city and everything moved so fast.

She kept thinking of Nova. She still needed to pick a grave memorial. She thought the marker should include a 3-D of a sun that had gone nova and the saying: "Your light will burn brightly throughout eternity." She also wanted to compile some video of key moments of the departed's life, a celebration of that life.

She ached to find a way to get her people back home. But that was one of the forbidden topics. It was not the right time to bring up the subject while the locals were still awaiting confirmation of how the aliens' war had turned out. True, the star people now had a comprehensive data base about the other species but it remained to be seen how much they would continue to be a threat.

She was thinking about all of these things as she prepared for her next interview, just moments away. She was sitting on the hover couch in her living room, wearing a skirt suit and a blouse. She fiddled with her hair and waved her hand like the locals had shown her, creating a mirror a couple feet from her face. She waved her hand again to magnify and checked out her hair, face and teeth. She snapped her fingers and the mirror vanished. At exactly the appointed time down to the nanosecond, a 3-D of a woman behind a desk appeared a few feet away.

"Hello, everyone," the image said, "This is Mindee Ferguson. I'm speaking with Zama Elle, former chiefexec of the planet A'laama in the Eta Cephai system. She's one of our allies who fought for humanity on the alien planet and one of the survivors of the vessel that was shipwrecked in their star system. Good morning, Zama. I'd like to personally thank you for fighting to protect and defend our fellow humans, a battle in which you were wounded. How have you been doing since arriving here?"

"Thank you, Mindee," she replied. "I'm doing well. I'm grateful for how generous everyone has been. We appreciate all that's been given to us: money, living quarters, food, offers for job training. I'd like to thank my . . . dear friend, Captain Erik Houston for coordinating the efforts to help us. Those of us who survived the A'laaman shipwreck escaped with virtually no possessions except the clothes we were wearing. I spent over three months with only one set of clothing. Now, thanks to your people I have a closet full of clothes and that alone makes me feel tremendously blessed."

The interview lasted about ten minutes. It seemed to go well and Zama was careful not to wander into any questionable subjects. The interviewer had treated her decently but the interviewee sighed with relief when it was over and the 3-D of the reporter disappeared. The A'laaman sat there for a few minutes, lost her in her thoughts. Her v-screen startled her when it blipped with three new messages. She pulled up each one. Three different strangers were each asking her on a date. She bit her lip, wondering how they had gotten through the security apps to get her private comm code.

Two new zings and a glimmer came in. One was a request for yet another interview. The next one looked like some kind of advertisement. The third message was a request for her to give a speech.

She arched her brow. Hmm. Interesting. But then she pictured her prepared text being blown to neutrinos by some government big wigs. She frowned.

"Ms. Elle, you have a visitor," said the automated door.

She jumped in her seat. "Who is it?"

"Captain Houston."

"Let him in, please."

She arose and walked toward the door.

"Good to see you," he said.

She gave him a hug. They walked to the hover couch and each sat down.

"Great interview," he said.

She nodded. "Things have been really busy since we've been back."

"Non-stop. And . . . I've been needing to talk with you."

She shot him an annoyed look. "What, more media guidelines?"

He chuckled. "No, I'm not here to *nano manage* you. This is personal."

"Go on, captain."

"It's about you and me."

She looked away. "Yes. You and me. What happens with you now that you're back home?"

"I'll have a couple months of shore leave. Time for R &R. Plenty of time to spend with you."

"Then what?"

"Well . . . unless I have any . . . *other commitments*, they'll probably send me back into space."

"I see."

"But . . . about us . . ."

She turned back toward him. Their eyes met. "I don't want to hurt you, but right now I'm not sure how I feel. When we were on your ship, you and I grew really close. I felt like I was falling in love with you all over again. But . . . we've gone through a lot. All the terrible things that happened on the return trip and it's been chaotic since then. I'm kinda overwhelmed."

He placed his hand on her arm. "What can I do to help?"

"For now, nothing. I just need some time. I've blocked out a few days to catch my breath. And think."

"I understand." He stood up. "Should I leave you alone?"

She nodded, her bangs falling into her eyes.

"Can I call you after that?"

"I'll call you."

He turned to leave.

"Erik," she called after him. He turned around. She walked over and hugged him. "Thank you."

He smiled.

After he left, she sat back down and returned to her thoughts. Then she spotted something on the floor. She leaned over and reached across the carpet. Her hand touched a grey, velvet box. She opened the lid. The light from the chandelier caught the facets of the most beautiful gemstone ring she had ever seen.

Erik tried zinging her again. No answer. Four days without talking with her was more than he could take. He summoned an auto cab and flew over to her building. The vehicle hovered beside the gated patio that would allow him to step out of the cab to within a few yards of her door, which was more than a hundred stories above ground level in this apartment tower. He strode over to the door.

"Hello, Captain Houston," the automated door greeted him.

"Hi. Is Ms. Elle in?"

"No, sir. I'm sorry, she's out."

He wrinkled his brow. "Where did she go?"

"She didn't say."

"Did she say when she's returning?"

"No, sir."

He turned to leave. He tried reaching her several more times that day but there was no answer. And she hadn't returned any of his messages.

42

Lost Forever?

Doctors wondered if Luci Strong would ever be the same. She had fully recovered physically and looked as good as ever. The regrown flesh on her face, arm and eye socket held no trace of scar tissue. But her mind was a twisted maze of alien and human thought patterns and emotions.

She had seemed normal enough for the first several days after returning to her ship. But then the worm had begun to take over. Once she had gotten back to regional base, she had remained in isolation. Specialists had analyzed her and nanomapped her brain. When she was conscious, she was violent, making it impossible to run tests on her. When she was out, some of the readouts were so bizarre that the doctors sought permission to study parts of the classified alien database to try to make sense of Luci's condition. The bug continued to run rampant in her mind as the input from the other species began to dominate, suppressing her humanity.

Finally, a Dr. Simoa was able to isolate the worm and deprogram the intrusion from the patient's mind. As soon as the foreign influence was gone, she went into convulsions then lapsed into a coma. Her vital signs declined and at one point it appeared she wouldn't make it. But after several hours she stabilized.

After she had been in a coma for a few days, the med and psych personnel summoned the other ship's officers to her side. Erik, Montoya, Irv and Marji each took turns talking to her and holding her hand. The captain told her how much he appreciated her bravery and how she had remained strong despite the alien's attacks. Montoya thanked her for being a loyal officer who had contributed much to the mission. Marji placed a vase with flowers on a table near the other lady. She told her how pretty she looked and how much she had valued her friendship over the years. The corners of Luci's mouth appeared to turn up ever so slightly. Irv, the ship's chaplain led the others in a prayer over their comrade. Her lips appeared to move slightly in response to his words.

Two more days passed. The parade of well wishers grew to include some of the other crew members. Whenever visitors left the room, they would frequently be accosted by human or android reporters seeking an update on Lt. Strong's condition.

That afternoon with the officers present, Luci opened her eyes. They grew wide as she saw the others staring down at her as she lay covered with a blanket. She looked alert. She reached toward Marji, who gave her a hug. Each of the others took a turn doing the same.

"You look good," said Montoya.

"I feel like hell," she replied.

"Welcome back," said the captain.

Her eyes narrowed. "Are we still in space?"

Marji shook her head. "We're home. We completed the mission and you're a big hero."

"You've been awarded the Spiral Galaxy medal and a platinum battle star," said Houston. "Headquarters will have a special ceremony once the docs have cleared you for discharge."

"Hero, huh? I'm just glad that junk in my mind is gone."

"Anything we can get you?" asked Marji.

"A frozen redberry twist, low cal. I've been getting fat in this sick bay. I need to get to a workout room. And Marjibelle, we still need to set up that double date. I promise not to show up as an alien."

43

Escape

S HE WALKED ALONG THE beach, the wind blowing her hair across the face. She had created a long, meandering line of footprints in the sand, a line that reached all the way back to the small, white cottage behind a dune hundreds of yards away. The surf lapped at her bare feet, the chilly water sending shivers up her spine. Her hair was wet and starting to frizz. She licked the salt spray from her lips. Some large cotton ball clouds broke the blueness of the sky. She pulled the water bottle from her shorts pocket and took a swig. She wiped her lips with her hand. She reached down and picked up a sea shell, adding it to the collection in the small bag that was tied to a belt loop. Overhead, a sea bird cried. No one else was around as far as they eye could see.

She looked further down the beach, her feet sinking a little into the moist sand. There was a dot off in the distance. As she continued her stroll, the dot became a figure. Another human 'way out here, miles from nowhere. As the other person grew closer, she squinted. Surely she was imagining this. She had to know for sure. She began to jog then broke into a run. The figure began to run as well. She was right, but how…? Her long legs rapidly closed the distance. She felt strong arms encircle her waist, hoist her up and swing her around. He continued to hold her in the air.

Wide-eyed, she looked down at his smiling face. "How did you ever find me?" she asked. He eased her down until her feet touched the sand.

"A long time ago, when you needed to get away, you had me take you to a beach in an uninhabited part of your planet."

"But there must be thousands of miles of beach on this planet. How . . .?"

He grinned even wider. "We can spot a specific grain of sand, a given sub-atomic particle from space. If you want to find someone badly enough you'll find them."

"I'm glad you found me." She reached for his hand and squeezed it.

"Is there someplace to sit down around here?"

"Do you see any furniture?"

He reached into his pocket, pulled out a light blue handkerchief and shook it. It opened up into a soft blanket. He held it over the sand and eased it to the ground. She smiled and sat down. He did likewise. "I need to show you something." He handed her an object the size and shape of an ancient pen. "Here, point this at that big cloud and push this button."

"Are you going to make it rain?" she laughed.

"No. This is some video shot through the gravitational lens at the outer reaches of this solar system. We've talked before about how each solar system has its own grav lens caused by gravity warping space. It's like a natural magnifying glass that allows astronomers to get some extreme closeups of a planet. You can pick up details just a few dozen yards across."

She nodded. "I remember. That's how you found out A'laama was inhabited so they sent your ship to visit us."

"That's right. We were shocked to find a colony when there was no record one had ever been started. Well, speaking of A'laama..."

A 3-D image projected out from the cloud. She gasped for breath. It was her planet as seen from space. "This is a recent live feed through the grav lens," he said. "Of course, this is how your planet looked five years ago due to the speed of light . . ."

"Yeah, yeah," she said, waving her hand because he was stating the obvious. "Do you know how old I am in planet years? I'm really getting up there."

"Another three and a half centuries or so and you'll catch up with me."

She dug her elbow into his ribs. "Old man."

"Old lady."

"Watch it, mister." She play-slapped him on the arm.

"Now, this should really get interesting for you," he said. He waved his hand and the shot zoomed in on the world's surface, on a certain continent, on a given geographic region. A hilly, jungle area and a large lake came into focus. The magnification continued to increase. She recognized the City of A'laama, the planet's capital and her former hometown. She felt her stomach drop. She could recognize many of the buildings, although the smaller ones looked a little fuzzy. This was almost as a close a look at the city as when she had floated over it with John in the Zep years earlier. Erik panned the view to look down on the mirror-walled Crystal Mansion, which had been her residence as chiefexec. The view swept over networks of criss-crossing lines and clusters of small dots. She assumed she was looking at networks of streets and houses out in the suburbs. She knew that somewhere down there were her children and grandkids. She saw a small comm tower, perhaps the audio and video scope her friend, Tanna had wanted to build. Right at this moment she might be tuning into broadcasts from other star systems. Zama got an overview of the new satellite towns, vast swatches of jungle, the Great Waterfall in the outback where the Batu tribe and other Non-techs lived. She ooh and awed at the new towns her people had established hundreds of miles from the capital, which had once been an isolated city-state. At the end of the tour, Zama's heart was pounding rapidly, her auto meds quickly responding before there was trouble.

"That was amazing," she said.

"There's more," he said. "I'd rather you stand for this."

"Okay."

He stood up and offered his hand to help her to her feet. He got down on one knee. "Zama Elle, I love you. I've always loved you. I've never stopped thinking about you since I left A'laama. Will you marry me?"

She felt faint. "I-I . . ."

"Take your time."

She eased back into a sitting position. He joined her. "This isn't easy," she said. She clasped both his hands and looked into his eyes. "I love you." Time seemed to stop. She had used the A'laaman word for love, not the starman word. Her word meant an intimate embrace, a deep union of two souls. Being completely at one with the other person. The look in his eyes made her certain he understood the meaning.

"I love you," she said, again using he A'Iaaman term. "I . . . was unable to admit that the last time I saw you."

"Why?"

She looked down at the sand. "I'm scared."

He squeezed her hands and waited until she finally met his gaze. His next words were so tender they moved her deep inside. "You're the one of the bravest women I know. Why are you afraid?"

She swallowed. Her mouth was dry. "I'm just wondering if our love is big enough. I'm an independent woman. I have strong opinions. That side of me has been suppressed lately due to your people's overseeing my every word."

He nodded. "We've been heavy-handed in dealing with you and your people. And I feel bad about that. It's not my idea and I'm only following orders. But soon the public and the media will find some other obsession. Then you can start building a more normal life."

"But for me normal means being outspoken and involved and calling the shots. I don't apologize for who I am. Can you and I really form a partnership for the rest of our lives? Do you really think we can make a marriage work?"

He stroked her face with the back of his hand. "You're spirited. Your own woman. Those are some of the things I love about you. I *don't know* how we'll make this work. But what I want most is to make you happy."

She swallowed hard and seemed incapable of speaking.

He continued: "You know what my career has been like . . ."

"I don't want to lose you," she said. "Not again." He put his arm around her. She laid her head on his shoulder.

For the next several minutes, the only sound was the gentle pounding of the surf. She sat up and looked out at the sea. She said: "The last time my life was at this big of a crossroads is the time you offered take me into the wilderness. I was afraid to leave my little cocoon and go explore the outback. But I was also curious to see what was there. And I definitely wanted to spend more time alone with you. If you hadn't pushed me to go, if you had just gone back into space and I'd remained content to stay in my own little city-state and not explore other parts of the world, if that had happened I never would have developed the boldness to build the starship and you and I would have never run into one other in that foreign solar system."

"So what are you saying?" he asked.

She pulled the bluestone ring out of her shorts pocket and slipped it onto her finger. His eyes looked like they would pop out of their sockets.

"Lose something, mister?" she asked.

Next it was her turn to be shocked. When she had retrieved the ring her hand had brushed against another object. She reached back into her pocket and pulled it out. It was a blank white tab. When she pressed on it a 3-D logo from a jewelry store, a price and a date appeared in mid-air. Tears formed in her eyes and began to roll down her face.

"W-what's wrong?" he asked.

"I thought you'd just bought this. But the date on this receipt is from years ago. Y-you bought this *before* you went on your mission. You loved me and you knew somehow we'd meet again."

"I didn't know. I hoped."

"You're a man of faith."

"Or else I'm crazy."

"A crazy man of faith."

"Or something."

"Shut up and kiss me," she said.

44

Reconciliation

THEY GOT MARRIED A week after he proposed in a small chapel on a tiny, isolated moon of an outer planet far from any prying e-media. The transparent walls looked out to the nearby world, which dominated the black sky. Both bride and groom were beaming.

Erik and his crew were all wearing their dress uniforms except for Luci and Marji, who wore bridesmaids dresses. Irv conducted the ceremony. When he pronounced the couple husband and wife, the room of star men and A'laamans erupted into loud cheering and clapping.

The reception involved a fusion of A'laaman and star man food, music and dance, with each group seeming to enjoy the others' culture. The diverse people mingled more that day than they had while living in close quarters aboard ship.

During the bridal dance, she held him close. Even while she treasured the moment, her mind flashed back to her conversation with Kelly all those years ago about the Prophecy and the part she played in fulfilling it. As her new husband and she glided around the dance floor, she thought of the Angel's prediction, at least the part that applied to her lifetime: "Before three hundred years have passed, three men shall come from the stars . . . .And after they arrive, your people shall never again be the same. In that day, there shall be an alliance between the leader of your seed and the leader from the stars. They shall be apart for a time but they shall come back together in the end."

She thought back to the adrenaline rush she had had when the star travelers had first landed on her planet. The landing party had included six people. There were three women, including Luci plus three men: the captain, Montoya and Minj. She recalled the time just after Erik had left A'laama and how the Angel's words about their eventually reuniting had lived in her heart. She had often gazed at the nighttime sky, wondering how and when she would see the star man again. Some nights she had even dreamed various scenarios about their reuniting. The passing of years had caused the promise to fade in her mind. Finally, she had begun to doubt. But after all this time she was back in the arms of her lover and they were committed to spending a lifetime together.

She held him closer. "Honey, remember when you were on my planet and we talked about the

Prophecy? And I teased you about how I never did tell you the last part of it?"

He pulled far enough away to look her in the eye. He wrinkled his brow. "Yeah, like it was some big secret. What's the story with that?"

She pulled him back close again and whispered in his ear. He pulled away again, his eyes looking very large. "Wha-at? You . . . me . . . we . . ."

She slowly nodded. "All foretold hundreds of years ago."

He stopped dancing. She took advantage of the moment and kissed him.

"And did they prophesy this?" he whispered, returning the kiss.

"Captain, I predict a lot of fun in your future," she said in a deep, seductive voice.

Later, the bride noticed Montoya gradually working his way closer to her. He got within a few feet then seemed to be looking in another direction. He began stroking his beard.

"Hi, Fred," she said.

He looked away.

She swallowed.

He cleared his throat. "I, uh . . . I owe you an apology."

She wrinkled her brow. "What do you mean?"

He looked down as he scraped the floor with the heel of one shoe. "I hate to admit this, but in the past I wasn't a big fan of yours."

Her stomach knotted. She'd often felt tense around Fred, although he had seemed to mellow toward her after their battle with the aliens. "Go on," she said.

"Women with strong personalities get on my nerves. Nothing against you. It's just the way I am. So, back on A'laama I used to make fun of you and call you names behind your back."

"Names?"

"Like Ice Queen and Ms. High Maintenance."

Her eyes narrowed. "So . . . I was the laughing stock of you and your crew."

"Just a few of the officers. None of the ladies. Just the guys."

She bit her lip. "You really don't need to tell me all this."

"I'm not saying this to hurt you. Please hear me out."

"Okay," she said, folding her arms.

"I've been wrong about you. Totally wrong. I haven't been a very good first officer for Erik, either. Haven't supported him and trusted him like I should. I've been disloyal, I . . ."

She laid her hand on his arm. "You're a good man. You may not like me but you saved my life. You didn't let me die on the shipwreck."

"I almost let you die on that planet."

She shook her head. "No, you didn't. Erik told me what happened. You were in the heat of battle and couldn't afford to leave your post. You took great care of me once the fighting was over. You kept us alive when the ship was failing. You're a kind and compassionate doctor."

"T-thank you," he mumbled. They were both silent for a long moment before he continued. "Even so, I've been hostile to you and I had no right to be. And the reason I acted that way . . . I-I hated that Erik was falling in love with you. He'd met various other women as we visited different planets and he'd had feelings for a couple of them. Erik and I . . . we've worked together a long time and even though I haven't been a perfect officer, I care about him. I didn't want him to get hurt again. Do you know how rare it is for a romance to work out between a career star jumper and a planet dweller?"

A sharp pain flashed through her chest then was gone again. "I-I'm sure it is. But you really can't call me a planet dweller anymore. I'm a *star lady* now."

"Yes. Yes, you are. You've earned that."

"Look, we're determined this will work. We have no doubts."

He smacked his forehead with his hand. "Man, I'm totally messing this up. I wasn't trying to say something bad about your relationship. I want to say something good."

She cocked her head.

"So Erik fell in love with you. He talked about you after we left your planet. And when he got back here to home base, all he could think about was you. And it made me mad. I thought he was crazy. When you showed up out there light years from nowhere, I resented it. But I've gotten to know you and you have a lot of character. You had enough guts to come planetside and risk your life fighting the aliens with us. I see how you and Erik look at one another and you make him very happy. Besides, it's no accident you two got back together. The chances were infinitesimal. Totally off the charts. Have you ever thought of how it couldn't possibly have happened by chance?"

Her smile lit up her entire face. "I know that more deeply than you could ever imagine."

The doctor smiled, too. "So if all the forces of heaven were determined to unite the two of you, who am I to speak badly of your love? Zama, you're a great wife for Erik. He's a very . . . a very blessed man."

Her throat closed up. She gave Fred a hug.

An autocam had recorded the ceremony and much of the reception then beamed the video to A'laama. Five years later the planet would receive an announcement telling the residents to prepare for a special broadcast. Then Zama's extended family plus Kelly, Tanna and all her other friends would see her wedding as if it were unfolding before them live. Zama still needed to send a more sobering message about the loss of A'laaman lives and the shipwreck in the alien solar system. But that story was for another day. Today was a day of joy.

After the couple spent a few days at the far end of the solar system they headed back to Newhattan. He helped her up the ramp of the *hovercab*. The cab lifted into the air and started off without creating any sensation of motion. It hurtled toward a group of pastel-colored skyscrapers that appeared to be made of light. The cab alighted at an outdoor landing pad halfway up one of the highest towers. The couple stepped out and entered the building through a soaring, arched doorway. Erik clasped her hand and beamed at her. Her face lit up as well. She squeezed his hand. He led her down a hallway. They stepped into a lift shaft that was so wide it dwarfed the one that had transported them from the alien planet back to the ship. She followed his lead when he hopped out of the shaft. He led her to a balcony. They were on the edge of an enormous atrium that appeared to be hundreds, no, thousands of feet high. She kept looking for the floor of this man-made canyon but never could see it although she nearly developed vertigo. She looked down to see people milling about on various floors below. She looked around and saw multi-dimensional sculptures that were each several stories high. Here and there various balconies jutted out to form grassy parks and wooded areas. Indoor rivers and streams spilled over man-made cliffs, forming waterfalls that created their own clouds of mist. There was even an indoor rainbow. Near the waterfalls, some of them tiered, scattered individual humans and small groups of them, looking like mites due to the sheer scale of their surroundings, flew to and fro. Some had wings attached to their backs and were soaring and gliding through the indoor sky. Fascinated, Zama felt like she had been here before. Then it hit her. This enchanting place had been in the video Erik has shown her advisors and she when he had visited her planet so long ago.

He asked her to go with him to rent a couple pairs of the wings. She nodded, still grinning. He approached a nearby stand and came back in a moment with two sets of giant appendages. He kept a pair of hawk's wings and handed her angel's wings. She smiled at him. He helped her attach the wings to her back. He leaped off a nearby platform then began soaring and gliding through the air. She swallowed hard and watched him for a moment. She jumped off the platform, adrenaline pumping through her. She soared through the sky and began to catch up with him. A natural athlete, it didn't take her long to gain proficiency with the wings. After gliding and dipping for a while, she began making loop-to-loops. She soared past a stone fountain that was several stories high and shaped like the head of a legendary creature called a lyon. The beast had a thick torrent of water pouring out of its mouth and into a pool far below. Numerous people the size of gnats were swimming and float-ing in the scallop-shaped pool. She began ducking in and out of various alcoves and spaces, Erik and she playing hide and seek with one another. He challenged her to a race but she won. He wanted to make it the best out of three and he won two. She zipped toward a far wall of the tower and viewed scenes that must be a couple hundred feet high made of mosaics of stained glass back lit by natural sunlight.

After a couple hours of flight, Zama and Erik alighted on the same balcony where they had begun their adventure. They turned in the wings and headed back to the mega liftshaft. They ascended for a couple minutes then exited. He led her down a plushy carpeted hallway and into a bar area. The bar looked out through some tall picture windows and outside the windows were . . . clouds! The sight disoriented her, causing her to grip his arm like a vice.

"Yes, those really are clouds out there," he said.

She shook her head.

While they had a drink he gazed into her eyes. "Before we went on our mission," he said, "I sat in this bar and had a drink and was looking out these same windows. Know what I was thinking about?"

"The mission?"

"You. I was thinking about you."

She leaned over and kissed him.

As they finished their drinks, he said: "You know, we can go out there."

"What? Walk on the clouds?"

He nodded.

She broke into a mischievous grin. "Let's do it."

He reached into a pack he was carrying and handed her a respirator. He donned a different one.

"Do we really need these?" she asked as memories from the ships' failing life support flashed back to her.

"Yes. The air is really thin this high up. And it's cold." He produced two handkerchiefs and shook first one, then the other. Each unfurled into a warmblanket. He wrapped one around him. She followed his example. They strolled over to the far wall and through a door. She unfastened her white sandals and carried them in one hand. The couple was now ankle deep in cloud. The frozen water vapor was cold on her feet. She took a few steps. "H-how is this possible? Force fields?"

He nodded.

She looked out across the carpet of cloud. In the distance, various other light-like skyscrapers poked through the cloudy floor. A shiver shook her shoulders and torso.

"It's best not to stay out too long," he said. "Let's go back inside."

"Aye aye, captain."

They returned indoors and stood in front of a two-story stone fireplace that hosted a roaring fire. A number of other cloud walkers were doing the same. About the time Zama was feeling toasty, she asked: "So when we do we get to see the bridal suite?"

His face turned red. He took off toward the lift shaft and she had to run to keep up.

"I thought we were already at the top of the building," she said.

He shook his head and pointed straight up. She craned her neck back. The shaft poked out beyond the building and continued upward until it disappeared. He popped into the shaft and she did the same. Zama looked down as she saw the pastel colored buildings and clouds shrink down to nothing then disappear. The sky outside the changed from medium blue to indigo to black. A few stars appeared. Then the sky looked like a million diamonds that had been poured out on black velvet. For a moment, she forgot to breathe.

The shaft ended and they stepped out onto a deep pile carpet. The outer walls of the building were transparent. She looked out at the blue-white planet below. She followed him down a hallway. He showed his hand print to the door, which eased open. He crouched down and picked her up in his arms.

She squealed with surprise.

She began to relax and placed her arm around his neck. She said: "Now, captain, you'd better not get me pregnant. I'm too old for that."

"I make no promises," said Erik.

The penthouse of the space hotel had a transparent roof and Zama was lingering there to gaze at the stars. Lost in her thoughts, she barely noticed when Erik, clad in a robe, put an arm around her waist. She absently grabbed his hand with her own and pulled him closer.

"You've overdressed for a honeymoon," he chuckled.

"So are you."

"I knew I'd find you here," he said.

"I feel closest to God when I'm looking at the stars," she replied, still gazing into space.

He nodded. "God is a lot bigger than any of us can grasp. See this map?" he said, waving his hand. A beach-ball sized black sphere freckled with hundreds of dots appeared in front of her, startling her away from the view out the window. Some dots in the sphere were large and bright. Others were dim enough to barely be visible. The dots spanned a virtual spectrum of color: white, blue-white, yellow, orange, red, brown. "These are all the stars within a fifty light year radius of Earth, the mother planet where everyone lived thousands of years ago," he said. "Within this fifty light year sphere is virtually all of human civilization. Now, watch this." He waved his hand and the vast the Milky Way galaxy replaced the first 3-D. "See this part of the spiral? This is called the Sagittarius Arm. Now, see this partial arm that branches off? This is called the Orion Spur."

"So where do we humans fit within all this?" she asked.

"See that tiny dot within the Orion Spur?"

She nodded.

"*That* represents all of human civilization."

She swallowed.

He continued: "Humanity has spent a couple thousand years scratching around in an area of space that's less than a pinprick compared with the size of the galaxy. Now, watch." He waved his hand again and the galaxy shrunk to a dot among myriad other galaxies then disappeared. "Four hundred billion galaxies with hundreds of billions of stars apiece. And God made all of that."

She nodded in awe.

They remained slient for a time, content to just hold one another. After a while she felt him grow tense. "I've been meaning to talk with you about something," he said. She felt the vibrations from his bass voice.

Her eyes seemed to grow larger. This sounded ominous.

He continued: "We've been so busy we haven't had a chance to talk about our future."

"W-what kind of future do you see for us, Captain Erik?"

"I've been star jumping a long time," he said. " If I left the service after all these years I'd get a big pension. And our government paid me a fortune to run that last mission. They knew there was a good chance I . . . I wouldn't come back." He looked deep into her eyes. "So I can afford to retire and have plenty of money for us to start a new life together. We can afford to do virtually anything you want. Even build a small ship so we can lead your people back to A'laama."

She burst into laughter.

He wrinkled his brow. "This is . . . funny?"

She gave him a long hug. "Erik, sweet Erik." she whispered in his ear. She pulled slightly away. "I'm anxious to get my people back home. And sure, I'd like to go back someday, too but I'm not in that big a hurry.

You really don't know me. That is, the *new* me. You're thinking of the old me, the woman you knew all those years ago."

He wrinkled his face as if this were some sort of female trick.

She placed her hands in his and squeezed. "Relax, honey." she said. "You've got all the time you need to get to know the new me. I'm not the same lady you knew before you took me into the wilderness, that gal who was afraid to leave her city-state. You, Erik Houston, single-handedly opened up a whole new world for me. After the terror and the adrenaline rush of that trip, I wanted more. I practiced with my first husband's ener pistol and learned to hunt. As we opened up new areas to explore on our planet, I toured the outback. I fought wild beasts when they attacked me. That's how I got this scar. The one you keep teasing me about," she said, hiking up her robe and revealing a long pink line on her right thigh.

She continued: "I climbed mountains, braved rivers. And it made me strong. When my heart was consumed with the desire to build a star-ship, I built it. When the stars beckoned, I went even though it almost killed me. Fighting those space creatures was the most frightening thing that ever happened to me, especially seeing them shoot at you. Then there were all the hazards on our return trip. You and I shared that oxygen mask and I've never felt closer to another human being. We could have died a half dozen times over. But we lived through it, you and I.

"Don't you *dare* retire, Erik Houston. Not for a long, long time. I've grown to love the things you love and it's made me love *you* more completely. After a couple months of mad, passionate love, I hope you're ready for your next mission. Because I'll be there by your side."

His face turned deep red. "I love you," he whispered in her ear. It was the A'laaman word. This was the first time he had ever said it to her. He swept her in his arms and kissed her with a passion that lit up the stars.

45

Epilogue

AFTER ERIK'S SHIP HAD left the alien star system, the war between the two rival navies continued for weeks until each side had suffered extreme losses, especially the outnumbered defenders. Humans and drone ships recorded the war and its aftermath from a safe distance by peering through the grav lenses of various star systems. Observers checking up on the alien planet in the aftermath of battle found a world surrounded by a thick layer of space debris, the shells and fragments of numerous ships from both sides of the war. The surface of the planet was in disarray with new craters that were miles across, canyon-sized fissures running hundreds of miles and massive volcanic activity and lava flows. Desert sand had melted under intense heat and eventually cooled to form seas of glass. The geography looked totally different than what Erik and his crew had seen. The forces unleashed against the planet had been so powerful that its orbit now had a pronounced wobble. As the scale of the devastation became known, it caused many humans to feel much more optimistic. But as a precaution most interstellar governments continued to expand and modernize their navies and step up security along the vast perimeter of civilization. They also dispatched ships to patrol just beyond the aliens' turf.

In the occasional far-flung solar system, ships would run into stragglers, two or three of the bug creatures' vessels that had remained separate from the main fleet. Most of the time, the humans were able to quickly dispatch their potential antagonists before they could do any harm. The insect people weren't so tough when they lacked superior numbers.

As the starship *Phoenix* eased into orbit around the rust-colored planet, Captain Bernard Eddleston thought about how much had happened in the twenty years since he had led the tiny exodus following the aliens' destruction of the Rantran III colony. Engineering crews had built the *Phoenix* in mere months, the first starman megaship constructed using the Minj-Bern Principle. Now that humans felt safe from the aliens, colonizing ships were once again going back into space. The first such interstellar ark had been dispatched to re-colonize Rantran III. As his crew made landing preparations, Eddleston thought of the history of some of the others who had accompanied him on the voyage.

Among the new colonists was Edd Brawnley, the astronomer who years ago had discovered the advancing alien fleet. The ships' chief engineer was his wife, Aida. The couple had been engaged prior to the firestorm that had vaporized the colony so many years ago. They had gotten married after they had left Rantran and reached the neighboring solar system. Their three teenage children accompanied them on the flight of the *Phoenix.*

Also onboard were several other settlers from the original colony. Then there was was twenty-year-old Matt Reddson. His mother had been pregnant with him at the time the aliens had invaded and she had fled the planet on one of the shuttles. He was anxious to see the orb that had been his mom's home. She had told him a number of stories about it.

While the colonists prepared for the landing, an advance party entered an ATV and rode a liftshaft to the planet surface. As the vehicle touched down, the shaft dissipated. Wearing warmsuits and respirators, the group set foot on the dusty surface. Their first act was to set up a small transmission tower. While they worked, the massive ship engaged multiple lift shafts to ferry personnel and equipment to their new home world. The immediate area began to bustle with activity as everyone worked to set up the pre-fab shelters that would serve as their temporary homes. Once the communication tower was operational, Eddleston ordered everyone to stop working. The sun was shining in the cloudless sky. The new residents all gathered around while the captain prepared to deliver a message technical personnel would be beam into space. As their excitement built, they broke into cheers for five minutes straight. Original colonists, the children and grandchildren of colonists, veteran space farers, retired old space jocks and recent Space Academy grads were all on hand to help re-start the colony. Years earlier, when the media had initially announced the mission, more than a hundred qualified people had volunteered for each available spot on the *Phoenix*.

Once the new arrivals to Rantran III finally settled down, the leader was able to start transmitting. "Greetings to all inhabited star systems and space stations," he said. "This is Bernie Eddleston, captain of the Starship *Phoenix*. I was also the captain of the *Pioneer*, the ship that brought the original colonists here over thirty Standard years ago. As of this date, we are founding a new colony of Rantran III with over four thousand settlers. The aliens hurt us but they couldn't destroy us. We're Rantran III and we're back!"

The crowd erupted into cheering once again. It was some time before they allowed the captain to continue but for the moment he was unable to speak further anyway.

www.ingramcontent.com/pod-product-compliance
Lightning Source LLC
Chambersburg PA
CBHW072356030726

47505CB00014B/1852

9 7 8 1 6 1 0 9 7 2 6 6 6